Molecular Computational Models:
Unconventional Approaches

Marian Gheorghe
University of Sheffield, UK

IDEA GROUP PUBLISHING
Hershey • London • Melbourne • Singapore

Acquisitions Editor:	Renée Davies
Development Editor:	Kristin Roth
Senior Managing Editor:	Amanda Appicello
Managing Editor:	Jennifer Neidig
Copy Editor:	Alison Smith
Typesetter:	Kristin Roth
Cover Design:	Lisa Tosheff
Printed at:	Yurchak Printing Inc.

Published in the United States of America by
    Idea Group Publishing (an imprint of Idea Group Inc.)
    701 E. Chocolate Avenue, Suite 200
    Hershey PA 17033
    Tel: 717-533-8845
    Fax: 717-533-8661
    E-mail: cust@idea-group.com
    Web site: http://www.idea-group.com

and in the United Kingdom by
    Idea Group Publishing (an imprint of Idea Group Inc.)
    3 Henrietta Street
    Covent Garden
    London WC2E 8LU
    Tel: 44 20 7240 0856
    Fax: 44 20 7379 3313
    Web site: http://www.eurospan.co.uk

Library of Congress Cataloging-in-Publication Data

Molecular computational models : unconventional approaches / Marian Gheorghe, editor.
    p. ; cm.
  Summary: "Molecular Computation Models: Unconventional Approaches is looking into new computational paradigms from both a theoretical perspective which offers a solid foundation of the models developed, as well as from a modeling angle, in order to reveal their effectiveness in modeling and simulating, especially biological systems. Tools and programming concepts and implementation issues are also discussed in the context of some experiments and comparative studies"--Provided by publisher.
  Includes bibliographical references.
  ISBN 1-59140-333-2 (hc) -- ISBN 1-59140-334-0 (sc)
  1. Molecular biology--Mathematical models.
  [DNLM: 1. Computational Biology. 2. Computers, Molecular. 3. Computer Simulation. 4. Models, Biological. 5. Models, Molecular. QU 26.5 M718 2005] I. Gheorghe, Marian, 1953-
  QH506.M66434 2005
  570'.1'1--dc22
                                          2004023592

eISBN 1-59140-335-9

British Cataloguing in Publication Data
A Cataloguing in Publication record for this book is available from the British Library.

All work contributed to this book is new, previously-unpublished material. The views expressed in this book are those of the authors, but not necessarily of the publisher.

Dedication

To the memory of Ray Paton, an eminent scholar, beloved friend, and loyal collaborator.

Molecular Computational Models:
Unconventional Approaches

Table of Contents

Preface

The interactions between different formal models and biology have a long-standing successful history. Results have been produced on both sides of this process while continuously new models are considered and new biological territories are uncovered.

On one hand, biological systems, as cells, have been recognised as complex systems requesting various models to simulate, explain, and verify their multiple properties. A very popular notion of complex systems is that of a very large number of simple and identical elements interacting to produce complex emergent behaviour. Unlike complex systems of simple elements, in which functions emerge from the properties of the network they form rather than from any specific elements, functions of biological systems rely on a combination of the network and the specific elements involved. Molecular biology has uncovered a great deal of biological facts, each one being very important in isolation, but various biological entities (cells, tissues, organs, organisms, colonies, societies) reveal very complex interactions. This level of aggregation, generally called systems biology, requires a combination of approaches, from experiments to abstract formal models able to manage the huge complexity of the biological systems to accurately represent and simulate them. In order to capture the complexity and the dynamicity of these systems, various models have been developed ranging from abstract mathematical-based models (either continuous or discrete approaches) to different specific programming paradigms or sophisticated software systems.

In the same time, natural sciences, and especially biology, represented a rich source of modelling paradigms. Well-defined areas of artificial intelligence (genetic algorithms, neural networks), mathematics, and theoretical computer science (L systems, DNA computing) are massively influenced by the behaviour of various biological entities and phenomena. In the last decades or so, new emerging fields of so-called "natural computing" identify new (unconventional) computational paradigms in different forms. There are attempts to define and investigate new mathematical or theoretical models inspired by nature, as well as investigations into defining programming paradigms that implement computational approaches suggested by biochemical phenomena. Especially since Adleman's experiment, these investigations received a new perspective. One hopes that global system-level behaviour may be translated into interactions of a myriad of components with simple behaviour and limited computing and communication capabilities that are able to express and solve, via various optimisations, complex problems otherwise hard to approach.

AIMS OF THE BOOK

Primarily, the book aims to:

- Overview a number of areas of natural computing (P systems, networks of evolutionary processors) by revealing the problems that are mostly researched and pointing towards future developments.
- Analyse and suggest various interactions between different approaches of the same biological phenomena.
- Discuss variants of classical concepts, such as dynamical systems, in rather unconventional settings (formal theory of languages, programming paradigms using evolutionary developments).
- Cover (nearly) all aspects of modelling (formal specifications, implementation, formal verification, simulations, predictions).
- Reveal the advantages of using hybrid complex unconventional models in order to express complex interactions occurring in complex biological entities (bacteria) or systems with a dynamic changeable structure.

AUDIENCE OF THE BOOK

The book is written to be used by a number of distinct audiences. It is mainly written for researchers developing new models, especially with a nature-inspired flavour, looking at new, surprising mathematical or theoretical properties of these models and also pursuing research in systems biology or computational biology. The book is also suitable for researchers and practitioners in the area of programming languages and paradigms who are interested in new programming concepts inspired from biology, with a bias towards developmental structures and topologies. Academics may find the book useful as a textbook presenting some advanced topics in computer science or a source of problems and research-led topics for projects or postgraduate courses.

OUTLINE OF THE BOOK

The book is organized into nine chapters.

Chapter 1 (Gheorghe Păun) presents an overview on membrane computing, a branch of natural computing whose initial goal was to abstract computing models from the structure and the functioning of the living cells. The research was initiated about five years ago (at the end of 1998), and since that time the subject has been developed significantly from a mathematical point of view. The basic types of results of this research concern computability power (in comparison with the standard Turing machines and their restrictions) and efficiency (the possibility to solve computationally hard problems, typically NP-complete problems, in a feasible time, almost polynomial). However, membrane computing has recently become attractive also as a framework for devising models of biological phenomena, with the tendency to provide tools for modelling the cell itself, not only the local processes. Chapter 1 surveys the basic elements of membrane computing, somewhat in its "historical" evolution, from biology to computer science and mathematics and back to biology. The presentation is informal, without any technical detail and an invitation to membrane computing intended to acquaint the nonmathematician reader with the main directions of research of the domain, the type of central results, and the possible lines of future development, including the possible interest of the biologist looking for discrete algorithmic tools for modelling cell phenomena.

Chapter 2 (Vincenzo Manca, Giuditta Franco, and Giuseppe Scollo) introduces some of the classical dynamics concepts in the basic mathematical setting of state transition systems where time and space are completely discrete and no structure is assumed on the state's space. Interesting relationships between attractors and recurrence are identified, and some features of chaos are expressed in simple, set theoretic terms. String dynamics is proposed as a unifying concept for dynamical systems arising from discrete models of computation together with illustrative examples. The relevance of state transition systems and string dynamics is discussed from the perspective of molecular computing.

Chapter 3 (Lila Kari, Elena Losseva, and Petr Sosík) is a survey that examines the question of managing errors that arise in DNA-based computation. Due to the inaccuracy of biochemical reactions, the experimental implementation of a DNA computation may lead to incorrectly calculated results. This chapter looks at different methods that can assist in reduction of such occurrences. The solutions to the problem of erroneous bio-computations are presented from the perspective of computer science techniques. Three main aspects of dealing with errors are covered: software simulations, algorithmic approaches, and theoretical methods. The objective of this survey is to explain how these tools can reduce errors associated with DNA computing.

Chapter 4 (Carlos Martín-Vide and Victor Mitrana) surveys, in a systematic and uniform way, the main results regarding different computational aspects of hybrid networks of evolutionary processors viewed both as generating and accepting devices, as well as of solving problems with these mechanisms. The chapter first explains how generating hybrid networks of evolutionary processors is computationally complete. The same computational power is reached by accepting hybrid networks of evolutionary processors. It is then defined as a computational complexity class of accepting hybrid networks of evolutionary processors and proven that this class equals the classical class NP. The chapter also presents a few NP-complete problems and recalls how they can be solved in linear time by accepting networks of evolutionary processors with linearly bounded resources (nodes, rules, symbols). Finally, the chapter discusses some possible directions for further research.

Chapter 5 (Andrés Cordón Franco, Miguel Angel Gutiérrez Naranjo, Mario J. Pérez Jiménez, and Agustín Riscos Nuñez) is devoted to the study of numerical NP-complete problems in the framework of cellular systems with membranes, also called P systems (Păun, 1998). The chapter presents *efficient* solutions to the Subset Sum and the Knapsack problems. These solutions are obtained via families of P systems with the capability of generating an exponential working space in polynomial time. A simulation tool for P systems, written in Prolog, is also described. As an illustration of the use of this tool, the chapter includes a session in the Prolog simulator implementing an algorithm to solve a subset sum or knapsack problem.

Chapter 6 (Jean-Louis Giavitto and Olivier Michel) focuses on a model inspired by biological development, both at the molecular and cellular levels. Such biological processes are particularly interesting for computer science because the dynamic organization emerges from many decentralized and local interactions that occur concurrently at several time and space scales. Thus, they provide a source of inspiration to solve various problems related to mobility, distributed systems, open systems, and others. The fundamental mechanisms of biological development are now understood as changes within a complex dynamical system. This chapter advocates that these fundamental mechanisms, although mainly developed in a continuous framework, can be rephrased in a discrete setting relying on the notion of rewriting in a topological setting. The discrete formulation is as formal as the continuous one, enables the simulation, and opens a way to the systematic study of the behavioral properties of the biological systems. Directly inspired from these developmental processes, it is presented as an experimental programming language called MGS. MGS is dedicated to the modelling and simulation of dynamical systems with dynamical structures. The chapter illustrates the basic notions of MGS through several algorithmic examples and by sketching various biological models.

Chapter 7 (Ray C. Paton, Richard Gregory, Costas Vlachos, JohnW. Palmer, Jon R. Saunders, and Q. H. Wu) describes two approaches to individual-based modelling that are based on bacterial evolution and bacterial ecologies. Some history of the individual-based modelling approach is presented and contrasted to traditional methods. Two related models of bacterial evolution are then discussed in some detail. The first model consists of populations of bacterial cells, each containing a genome, or gene products developed through transcription cascade and mutation. As a result, this model contains multiple time scales and is very fine-grained. The second model employs a coarser-grained, agent-based architecture, designed to explore the evolvability of adaptive behavioural strategies in artificial bacterial ecologies. The organisms in this approach are represented by mutating learning classifier systems. Finally, the subject of computability on parallel machines and clusters is applied to these models, with the aim of making them efficiently scalable to the point of being biologically realistic by containing sufficient numbers of complex individuals.

Chapter 8 (Gabriel Ciobanu) describes a model of the molecular networks by using a system of communicating automata as a dynamic structure and discrete event system, providing interesting theoretical results. This formal model provides a detailed approach of the biological system, and its implementation is able to handle large amounts of data. This model is applied to a T cell signalling network. A T cell shows a hierarchical organization depending on various factors. Some mechanisms are still unresolved, including contribution of each signalling pathway to each response type. The software tool produced is used to simulate and analyze T cell behaviour. The simulation reflects quite faithfully the process of T cell activation and T cell responses. This increases the confidence to use this model and its implementation both as a descriptive and prescriptive tool. The interactions that govern T cell behaviour are simulated and analyzed, providing statistical correlations according to software experiments, together with new insights on signalling networks that trigger immunological responses. The software tool allows users to systematically perturb and monitor the components of a T cell molecular network, capturing relevant biological information to immunology.

Chapter 9 (Petros Kefalas, George Eleftherakis, Mike Holcombe, and Ioanna Stamatopoulou) presents a formal method, namely X-machines, used to specify, verify, and test individual agents. Multi-agent systems are highly dynamic since the agents' abilities and the system configuration often changes over time. In some ways, such multi-agent systems seem to behave like biological processes; new agents appear in the system, some others cease to exist, and communication between agents changes. One of the challenges of multi-agent systems is to attempt to formally model their dynamic configuration. Utilized concepts from biological processes can identify and define a set of operations that are able to reconfigure a multi-agent system. This chapter presents an example of these concepts, in which a biology-inspired system is incrementally built in order to meet our objective.

Acknowledgments

The chapter authors are first acknowledged not only for their contributions to this book, but also for their efforts to review other chapters. For the reviewing process, special gratitude is also due to other colleagues and friends who kindly read and reviewed different chapters: T. Balanescu, D. Besozzi, E. Csuhaj-Varju, M. Margenstern, I. Petre, and G. Vaszil.

Particular thanks go to the publishing team of Idea Group Inc., in particular to Mehdi Khosrow-Pour, who persuaded me to accept the invitation to take on the editorial responsibility, and to Michele Rossi, who continuously helped me with requested details regarding this project and its schedule. Many thanks also to the staff of Idea Group Inc., who supported the project throughout its development.
To all these wonderful people, a deep bow.

Marian Gheorghe
University of Sheffield, United Kingdom

Chapter I

Membrane Computing:
Main Ideas, Basic Results, Applications

Gheorghe Păun

Institute of Mathematics of the Romanian Academy, Romania, and
Research Group on Natural Computing, University of Sevilla, Spain

ABSTRACT

Membrane computing is a branch of natural computing whose initial goal was to abstract computing models from the structure and the functioning of living cells. The research was initiated about five years ago (at the end of 1998), and since that time the area has been developed significantly from a mathematical point of view. The basic types of results of this research concern the computability power (in comparison with the standard Turing machines and their restrictions) and the efficiency (the possibility to solve computationally hard problems, typically NP-complete problems, in a feasible time and typically polynomial). However, membrane computing has recently become attractive also as a framework for devising models of biological phenomena, with the tendency to provide tools for modelling the cell itself, not only the local processes. This chapter surveys the basic elements of membrane computing, somewhat in its "historical" evolution:

from biology to computer science and mathematics and back to biology.
The presentation is informal, without any technical detail, and an invitation
to membrane computing intended to acquaint the nonmathematician
reader with the main directions of research of the domain, the type of
central results, and the possible lines of future development, including the
possible interest of the biologist looking for discrete algorithmic tools for
modelling cell phenomena.

INTRODUCTION

In some sense, the whole history of computer science is the history of
attempts to discover, study, and, if possible, implement computing ideas,
models, and paradigms the same way nature—humans included—computes.
We do not enter here into the debate whether or not the processes taking place
in nature are by themselves "computations", or whether we, *Homo sapiens*,
interpret them as computations. But we do recall that when defining the
computing model now known as the Turing machine, which provides the
standard—and by now definition—of what is computable, A. Turing (in 1935
and 1936) explicitly wanted to abstract and model what a clerk in a bank is
doing when computing with numbers. One decade later, McCullock, Pitts, and
Kleene founded the finite automata theory starting from modelling the neuron
and the neural nets; still later, this led to the area known now as neural
computing. Genetic algorithms and evolutionary computing and programming
are now well-established (and frequently applied) areas of computer science.
One decade ago, Adleman's history-making experiment of computing with
DNA molecules was reported, proving not only that biology can inspire
computer and algorithm design for electronic computers, but also that biological support (a bioware) can be used for computing. In recent years, the search
of computing ideas, models, and paradigms in biology, or in nature in general,
has become explicit and systematic under the general name of natural computing.

Membrane computing is a part of this intellectual enterprise, starting from
two general premises: 1) nature has evolved the living beings from the
biochemistry in the compartments of a cell, to tissues, organs, organisms,
populations, during billions of years, with goals different from those of computer science but which often turn out to be surprisingly useful for computing
and computer science (the best illustration is that of genetic algorithms and
evolutionary computing), and 2) The cell is the smallest living thing, and at the
same time it is a marvellous, tiny machinery with a complex structure, an

intricate inner activity, and an exquisite relationship with its environment — the neighbouring cells included.

The challenging issue is whether or not the structure and the functioning of the living cell can provide any suggestions to computer science. Membrane computing has emerged as a possible answer to this challenge, proposing a series of models (actually, a general framework for devising models) inspired by cell structure, or a compartmentalized space defined by a hierarchical arrangement of membranes, functioning, which is the biochemical processes taking place in the compartments of the membrane hierarchy and the way the compartments cooperate and communicate by passing chemicals and information across membranes, and cell organization in the tissue. These models, called P systems, were investigated as mathematical objects, with the main goals relating to computer science: computational power (in comparison with Turing machines and their restrictions), and usefulness in solving computationally hard problems. The field simply flourished at this level. Comprehensive information can be found in the Web page (organized under the auspices of the European Molecular Computing Consortium, EMCC) at *http://psystems.disco.unimib.it*; a 2002 presentation can be found in Gheorghe Păun's *Computing with Membranes* (2002).

Concomitantly with this strong mathematical development of membrane computing, two phenomena can be noticed. The first one concerns the general relationship of biology with computer science (we will elaborate on this also in the section below): although biological investigations have significantly benefited from computers and computability (the genome project is a perfect illustration), a breakthrough is still needed to model complex biological systems and the cell, as small as it is, is one of the biological systems which is still too complex for current information science and technology to be modelled as a whole. The second phenomenon is internal to membrane computing, and it is quite similar to phenomena of this type from other scientific areas that were successfully mathematized (physics and linguistics are good illustrations). When models born for describing "objects" from an area A become abstract enough, essentialized enough, then they can be used for describing "objects" from other areas, sometimes far from the initial area A. In particular, these models, even developed with completely different goals, can come back to the area that suggested them, possibly useful for applications, and possibly returning relevant findings. This is exactly the case of P systems, which, used abstractly and with substantial mathematical knowledge, were applied not only to biology, but also to linguistics, management, and specific computing areas (e.g., sorting and merging, computer graphics).

The present chapter will briefly touch on all of these issues, from describing classes of P systems with biological and mathematical motivation, to computability results, software implementations, and applications. The chapter will also mention what membrane computing is not yet — because many things remain to be done. Membrane computing itself is still in its infancy as a source of models for the use of the biologist.

DISCRETE VERSUS CONTINUOUS MODELLING IN BIOLOGY

Before entering into the details of the chapter's primary topic, we want to quickly elaborate on a debate of interest related to this topic: Which kind of mathematics, discrete or continuous, is the most suitable or useful for modelling biological systems and phenomena? Of course, stated in this way, the question does not make much sense, the answer is obvious: both kinds of mathematical models are (or could be) useful. However, the debate is not senseless. Actually, it has not appeared in biology, but was raised in linguistics several decades ago (see details in Marcus, 2004). The point is that, since Newton at least, the continuous mathematics, primarily calculus and differential equations, proved to be extremely useful for physics, astronomy, engineering, and other fields. The whole relativity can be considered a sort of differential geometry of space and time. Under this strong impression, the belief has appeared that the same type of approach and the same tools as used in physics will be useful in many other domains, such as sociology, linguistics, and, our case of interest, biology.

However, on one hand, it was realized soon (and explicitly advocated for linguistics) that the genuine nature of many phenomena are of a discrete type. In particular, biochemical processes can be described by differential equations, but this is in many cases just an approximation of discrete by continuous, in many cases simultaneous with an approximation of finite by infinite. Moreover, the systems of differential equations corresponding to nontrivial biological phenomena turned out to be difficult to handle as the biological processes are complex in themselves with nonlinear and nondeterministic behaviours, small changes in the process lead to major changes of equations, the compartmentalization cannot be captured, scalability raises serious problems, and differential equations are difficult for biologists to understand.

In mirror with these observations, we have to mention the huge progresses of computer science, which made possible not only the accumulation of immense databases of biological information, but also the possibility to handle these databases. Then algorithmic mathematics came onto the stage, having also the instruments to handle the algorithms — the continuously improved computers. However, computers are in a direct way related to discrete mathematics (starting with the reduction of any information to a 0/1 representation) and to symbolic computing.

In short, the idea of using discrete and algorithmic models in biology is more and more supported, and the case of linguistics is rather encouraging. The specific task, sometimes mentioned as the challenge of the 21st century, or as the postgenomic challenge for bio-informatics — that of modelling the whole cell and simulating it on a computer — seems to be reasonably addressed mainly by involving discrete (algorithmic) mathematics. This does not exclude continuous mathematics, which is adequate (and relevant) for describing local quantitative processes, but pleads for global models of the cell of a "grammatical" type — the reactions taking place in the compartments of a cell can be interpreted as "rewriting rules", processing the "chemical-words" swimming in solution in the respective compartments.

This discussion is related to several other distinctions relevant for the "mathematical biology", a discipline which seems to have borne: quantitative versus qualitative investigations, experimental versus model-intermediated research (with the complex relationship between models and experiments), formal versus informal approaches, explanatory versus predictive models, structure versus behaviour (and the interplay of the two), and top-down versus bottom-up approaches. We do not enter this more general debate about what type of modelling is appropriate (this is the title of a paragraph from the introduction of Bower and Bolouri, eds., 2001), but we conclude with the observation that, although the previous reasoning can be considered as a plead *pro domo sua* for discrete algorithmic modelling, we believe in it, together with other authors, namely Harel (2004), Holcombe (2001), Kitano (2002), Tomita (2001) — with a title which became a slogan of researches related to cell simulation, and Webb and White (2004).

THE BASIC CLASS OF P SYSTEMS

We introduce now the fundamental ideas of membrane computing, mainly as suggested by the biology (the structure and functioning) of the cell. What we look for is a computing device, and to this aim we need data structures, operations with these data structures, an architecture of our "computer", a systematic manner to define computations, and results of computations. A first answer to all of these issues is given below; later, further biologically or mathematically inspired features will be added.

The fundamental ingredients of a membrane system (we use the standard name of "P system") are 1) the *membrane structure* and 2) the sets of *evolution rules* that process 3) *multisets* of 4) *objects* placed in the compartments of the membrane structure.

A membrane structure is a hierarchically arranged set of membranes, as suggested in Figure 1, where we distinguish the external membrane (corresponding to the plasma membrane and usually called the "skin" membrane) and several internal membranes (corresponding to the membranes present in a cell, around the nucleus, in the Golgi apparatus, vesicles, mitochondria, etc.); a membrane without any other membrane inside it is said to be *elementary*. Each membrane determines a compartment, also called region, which is the space delimited from above by it and from below by the membranes placed directly inside, if any exists. The correspondence membrane region is one-to-one, which is why we sometimes use these terms interchangeably; also, we identify by the same label a membrane and its associated region.

In the basic variant of P systems, each region contains a multiset of symbol-objects, which correspond to the chemicals swimming in a solution in a cell compartment; these chemicals are considered here as unstructured, which is why we describe them by symbols from a given alphabet.

The objects evolve by means of evolution rules, which are also localized, associated with the regions of the membrane structure. The rules correspond to the chemical reactions possible in the compartments of a cell. The typical form of such a rule is $aad \rightarrow (a,here)(b,out)(b,in)$, with the following meaning: two copies of object a and one copy of object b react (and they are consumed

Figure 1. A membrane structure

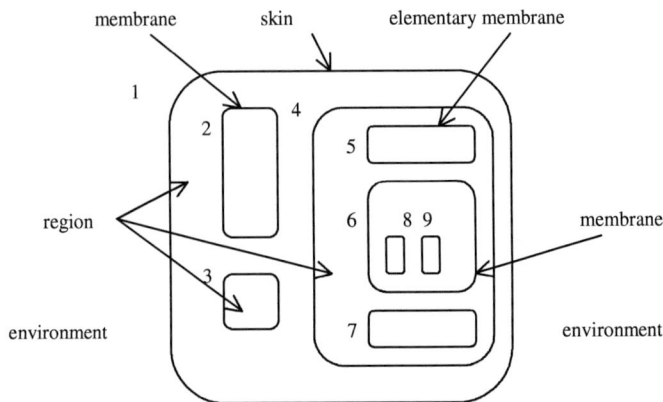

during this reaction) and the reaction produces one copy of *a* and two copies of *b*. The new copy of *a* remains in the same region (indication *here*), one of the copies of *b* exits the compartment and goes to the surrounding region (indication *out*), and the other enters one of the directly inner membranes (indication *in*). We say that the objects *a, b, b* are *communicated* as indicated by the commands associated with them in the right hand member of the rule. When an object exits a compartment, it will go to the surrounding compartment; in the case of the skin membrane this is the environment, hence the object is "lost", it cannot be brought back into the system by rules such as those just mentioned. If no inner membrane exists (that is, the rule is associated with an elementary membrane), then the indication *in* cannot be followed, and the rule cannot be applied.

A rule as that just discussed, with at least two objects in its left hand side, is said to be "cooperative". A particular case is that of catalytic rules, of the form $ca \rightarrow cv$, where *c* is an object (called catalyst) that assists the object *a* to evolve into the multiset *v*. Rules of the form $a \rightarrow v$, where *a* is an object, are called "noncooperative".

The rules can also have the form $u \rightarrow v\delta$, where δ denotes the action of dissolving the membrane — if the rule is applied, then the respective membrane disappears, and its contents, objects, and membranes alike are left free in the surrounding membrane. The rules of the dissolved membrane are removed at the same time with the membrane. The skin membrane is never dissolved.

The communication of objects through membranes reminds us that the biological membranes contain various protein channels through which the molecules can pass (in a passive way due to concentration difference, or in an active way with a consumption of energy), in a rather selective manner. The fact that the communication of objects from one compartment to a neighbouring compartment is controlled by the "reaction rules" is mathematically attractive but not quite realistic from a biological point of view, which is why variants were also considered in which the two processes are separated. In this case, the evolution is controlled by rules like those discussed earlier, without target indications, and the communication is controlled by specific rules (symport/antiport rules as described later in this chapter).

Note that evolution rules are stated in terms of names of objects, they are "multiset rewriting rules", while their application or execution is accomplished using copies of objects. The data structure we work with is the multiset of objects, or sets with multiplicities associated with their elements.

A membrane structure and the multisets of objects from its compartments identify a configuration of a P system. The initial configuration is given by specifying the membrane structure and the multisets of objects available in its compartments at the beginning of a computation.

Membrane systems are synchronous devices, in the sense that a global clock is assumed, that is, the same clock marks the time for all regions of the system. In each time unit a transformation of a configuration of the system — we call it transition — takes place by applying the rules in each region, in a nondeterministic and maximally parallel manner. This means that the objects to evolve and the rules governing this evolution are chosen in a nondeterministic way; this choice is "exhaustive" in the sense that, after the choice is made, no rule can be applied in the same evolution step to the remaining objects.

It is instructive to see a transition in a P system as a "macrostep" consisting of several "microsteps" performed after each other. Consider a region r of the system. First, we assign occurrences of objects from r to rules from r, nondeterministically choosing rules and objects until no further assignment is possible (note that the multiplicity of objects present in r is crucial in this microstep). Then, all these assigned objects are removed from the current multiset of objects in r, and occurrences of all objects specified by the right hand sides of the chosen rules are added to this multiset, together with their transfer commands *in, out,* and *here*. Now, all transfers indicated by commands *in* and *out* are executed, and copies of objects with the transfer command *here* remain in region r. Finally, the communication commands are removed, and a macrostep is completed for r. Since all regions are processed simultaneously (with all microsteps performed synchronously), this completes the global macrostep.

Of course, the previous way of using the rules from the regions of a P system prompts the nondeterminism and the partial parallelism from cell compartments, with the observation that the maximal parallelism is mathematically oriented (which is rather useful in proofs). When using P systems as biological models, this feature should be replaced with more realistic features (e.g., reaction rates, probabilities, partial parallelism).

A sequence of transitions constitutes a computation. A computation is successful if it halts, or reaches a configuration where no rule can be applied to the existing objects. With a halting computation we can associate a result in various ways. The simplest possibility is to count the objects present in the halting configuration in a specified elementary membrane; this is called internal output. We can also count the objects that leave the system during the computation, called "external output". In both cases, the result is a number. If we distinguish among different objects, then we can have, as the result, a vector of natural numbers. The objects that leave the system can also be arranged in a sequence according to the moments when they exit the skin membrane, and in this case, the result is a string.

Because of the nondeterminism of the application of rules, starting from an initial configuration, we can get several successful computations, hence several results. Thus, a P system computes (one also used to say "generates") a set of numbers, or a set of vectors of numbers, or a language, depending on the way the output is defined. The case of language is important for the qualitative difference between the "loose" data structure we use inside the system (vectors of numbers) and the data structure of the resulting strings, in which we also have a "syntax", or positional information.

Consequently, P systems are distributed systems with a highly parallel behaviour (besides the parallel processing of objects in each region, all regions simultaneously evolve their contents), processing multisets of objects in a synchronous manner.

We do not give here a formal definition of a P system. The reader interested in mathematical details, in rigorous definitions, or in further bibliographical information can consult the mentioned monograph by Păun (2002), the introductory paper by Păun and Rozenberg (2002), as well as the relevant papers from the Web bibliography mentioned in Section 1 of this chapter. The collective volumes Alhazov et al. (2003), Calude et al. (2001), Cavaliere et al. (2003), Martín-Vide et al. (2004), and Păun et al. (2003) are of particular interest since they contain both theoretical developments and applications. Of course, when presenting a P system we have to specify: the alphabet of objects (usually a finite, nonempty alphabet of abstract symbols identifying the objects), the membrane structure (the usual description of the tree associated with the membrane structure is represented by a labelled tree, by an Euler-Venn diagram like in Figure 1, or, more compactly, by a string of labelled matching parentheses), the multisets of objects present in each region of the system (represented in any suitable manner, such as by strings of symbol-objects, with the number of occurrences of a symbol in a string being the multiplicity of the object identified by that symbol in the multiset represented by the considered string), the sets of evolution rules associated with each region, as well as the indication of the way the output is defined (internally or externally, as a number or a string; in the internal mode of defining the result of a computation, we have to specify the elementary membrane where the objects should be counted at the end of halting computations).

Graphically, a P system can be represented in a natural and suggestive way, as an Euler-Venn diagram with the multisets of objects and the rules specified in each region. Figure 2 gives an example from Păun (2002) of a P system that computes the squares of natural non-null numbers (the output is read in membrane 1, which should be elementary at the end of a computation).

Because we want to have here a nontrivial example, we also use an ingredient — a *priority* relation among rules — that extends the previous definition. This extension is given by means of a partial order relation on the set of rules from each region. In the presence of such a relation, a rule can be applied in a given step only if no rule of a higher priority is applicable in the same region. This corresponds to the biological fact that there are reactions which are more active than others.

The computation in the system from Figure 2 develops as follows. In the central membrane, that with the label 3, for $n \geq 0$ times the rule $a \rightarrow ab$ is applied in parallel with $f \rightarrow ff$ (in this way, the number of bs grows each step by one, while the number of fs is doubled in each step), followed by the rule $a \rightarrow b\delta$ (again in parallel with $f \rightarrow ff$). The membrane is dissolved, and its contents ($n+1$ copies of b, and 2^{n+1} copies of f) are left free in membrane 2, which now can start using its rules. In the next step all objects b become d, while the number of copies of f is divided by 2 by using the rule $ff \rightarrow f$ (it has priority over the rule $f \rightarrow d\delta$). Then, in each step each d produces one copy of e, while the number of fs is divided by 2. This process can continue until we get a configuration with only one copy of f present; at that step we have to use the rule $f \rightarrow d\delta$ (the rule $ff \rightarrow f$ is no longer applicable). Hence, membrane 2 is also dissolved. Because we have applied the rule $d \rightarrow de$, in parallel for all copies of d (there are $n+1$

Figure 2. The initial configuration of a P system (with rules included)

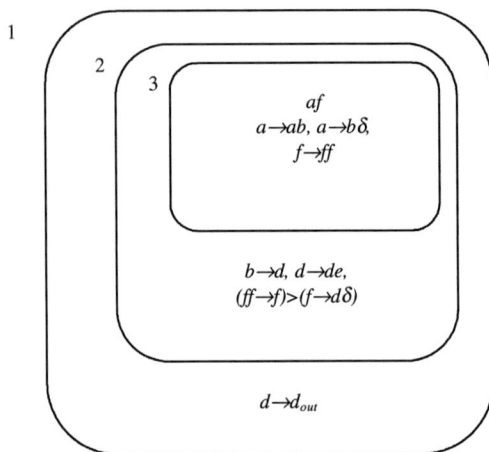

such copies), during $n+1$ steps, we have $(n+1)(n+1)$ copies of e and $n+2$ copies of d (one of them was produced by the rule $f \rightarrow d\delta$) present in the skin membrane of the system (the unique membrane still present). The objects d are removed from the system, and the computation halts, as no rule is available in region 1 for processing the object e.

SOME MODIFICATIONS AND EXTENSIONS

Many modifications and extensions of the very basic model we have sketched are discussed in the literature. We will briefly mention here only a few of them.

The motivation for such modifications and extensions comes from various directions: the attempt to capture further biological features, both to get a model as "realistic" as possible, and in the hope to get a computationally better model; the attempt to have as powerful as possible computing models (for computing models equal in power with Turing machines, one often obtains universality results in a direct way, which, from a practical computer science point of view, means programmability); the need to have as efficient as possible computing models (able to solve hard problems in a feasible time); the desire to have mathematically elegant models (minimal, without involving "too many" ingredients). Needless to say, these goals form a contradictory set, improving in some of them, and in general means losing in others, which explains why so many types of membrane systems were considered. This can be related to the general trade-off between universality and programmability, efficiency, and evolvability or learnability, as documented by Conrad (1998).

An extension was already mentioned in the previous section — it is a priority relation among rules, and very useful in programming the computations in a P system.

Another useful "control device" is the possibility to modify the membrane permeability. Thus, a membrane can be made thinner (action δ) or thicker (action τ). These actions are associated with evolution rules, which can be of the form $u \rightarrow v\delta$ or $u \rightarrow v\tau$. Using such a rule means using $u \rightarrow v$ in the standard mode, and then to applying action δ or τ. The interplay between the actions δ and τ briefly described earlier is pictorially described in Figure 3. A membrane of normal thickness (indicated in the figure by NT) is dissolved by action δ (the objects of a dissolved membrane remain in the region surrounding it, while the rules are removed; the skin membrane cannot be dissolved), or made impermeable (no object can pass through such a membrane) by action τ. An impermeable membrane (indicated in Figure 3 by IM) is returned to normal

Figure 3. The interplay of actions δ and τ

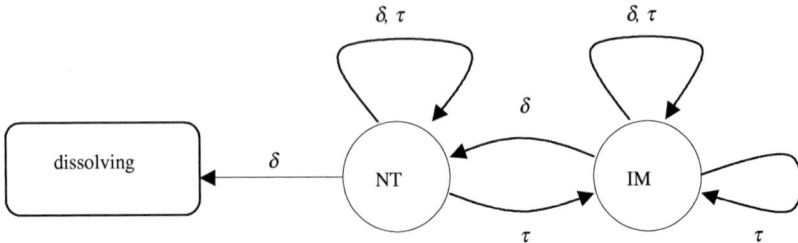

thickness (hence it is again permeable) by action δ. If δ and τ are introduced simultaneously, then the thickness of the membrane is not changed.

Many possibilities are offered by the communication commands. In the variant mentioned in the previous section, the target indications were *here, in,* and *out*, but there are various other possibilities. For instance, we can consider stronger target indications, of the form *in$_j$*, which specify the label of the lower level membrane where an object should be transferred. Another possibility is to associate electrical charges both with objects and with membranes. A polarized object will enter the region of any adjacently lower membrane of the opposite polarization; the polarization of objects and of membranes may change during the computation.

Several, more elaborated possibilities are offered by handling the rules. In the standard setup described earlier in this chapter, the rules are used in a maximally parallel manner, with the objects and the rules chosen nondeterministically. It is, however, natural to also use the rules in a sequential manner, one in each region in each transition. Then, the rules can also be moved through regions, in the same way as the objects are moved. Finally, the rules can also be consumed when applied, and new rules can be introduced at the same time. The rules are then of the form $r: u \rightarrow v/R$, where r is a label, $u \rightarrow v$ is a usual multiset processing rule, and R is a set of labels of rules. After applying the rule $r: u \rightarrow v$, this rule is no longer available, but the rules indicated by R do become available). This is a counterpart of the biological fact that many reactions taking place in a cell are protein-driven, and the number of (copies of) proteins matters.

COMPUTING BY COMMUNICATION

An important class of P systems is that of symport/antiport systems, where the whole computation is performed by moving objects across membranes, based on operations directly inspired from biology.

In the systems presented in the previous sections, the symbol-objects were processed by multiset rewriting-like rules (some objects are transformed into other objects, which have associated communication targets). Coming closer to the transmembrane transfer of molecules, we can consider purely communicative systems, based on the three classes of such transfer known in the biology of membranes: *uniport, symport,* and *antiport* (see Alberts et al. (2002) for details). Symport refers to the transport where two or more molecules pass together through a membrane in the same direction, antiport refers to the transport where two or more molecules pass through a membrane simultaneously, but in opposite directions, while uniport is when a molecule does not need a "partner" for a passage.

Figure 4 illustrates these ideas. In the case of promoted transport, a specific protein — indicated by C in the figure — should be bound to the membrane in the vicinity of the protein channel.

In terms of P systems, we can consider object processing rules of the following forms: a symport rule (associated with a membrane i) is of the form (ab,in) or (ab,out), stating that the objects a and b enter and exit together with membrane i, while an antiport rule is of the form $(a,out;b,in)$, stating that a exits

Figure 4. Symport, antiport, and promoted symport

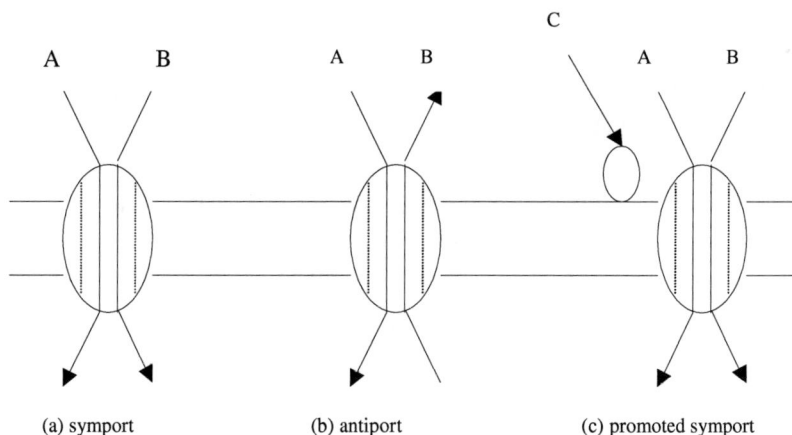

(a) symport   (b) antiport   (c) promoted symport

and *b* enters membrane *i* simultaneously; uniport corresponds to a particular case of symport rules, of the form *(a,in)* and *(a,out)*.

A direct generalization is to use rules of the forms *(x,in)*, *(x,out)*, and *(x,out;y,in)*, where *x* and *y* are arbitrary multisets of objects (the size of these multisets is called the "weight" of the respective rules). Note that in the case when we use symport or antiport rules associated with the skin membrane we can send objects out of the system and we can bring into the system objects from the environment.

A P system with symport/antiport rules has the same architecture as a system with multiset rewriting rules: alphabet of objects, membrane structure, initial multisets in the regions of the membrane structure, sets of rules associated with the membranes, and possibly an output membrane — with one additional component, the set of objects present in the environment. If an object is present in the environment at the beginning of a computation, then it is considered available in arbitrarily many copies (the environment is inexhaustible). This is an important detail; because by communication we do not create new objects, we need a supplier of objects in the environment. Otherwise, we are only able to handle a finite population of objects — those provided in the initial multiset.

The functioning of a P system with symport/antiport rules is the same as for systems with multiset rewriting rules: the transition from a configuration to another configuration is done by applying the rules in a nondeterministic maximally parallel manner, to the objects available in the regions of the system and in the environment, as requested by the used rules. When a halting configuration is reached, we get a result, in a specified output membrane (the environment already contains objects, which cannot be used in a trivial way for collecting the result).

Another way to organize computations by communication only is to use carriers (corresponding to "vectors" and to plasmids used in biochemistry), that is, to consider objects of two types: carriers ("vehicles") and passengers. As in the case of symport/antiport, no object ever changes. The passengers can pass through membranes only when carried by carriers. The used rules specify the way of attaching passengers to carriers, the way these "aggregates" pass through membranes, and the way to detach passengers from carriers.

In the case of symport/antiport and in the case of carriers, no object is created or destroyed, only the location of the objects can be changed. Hence, the conservation law is observed — which does not necessarily happen in other classes of P systems. Also, in both of these cases the environment is an active participant in the computation, holding as many copies of each object as necessary, and involved in a two-way communication with the skin region of the system.

In the case of systems with symport/antiport rules, one can associate a string with a computation by considering the trace of a specified object — the "traveller" — through membranes (the sequence of labels of membranes visited by the traveller during a successful computation). Hence, again a language is associated with a device working with numbers as the internal data structure.

The symport/antiport rules can be used also for defining a class of P systems where the evolution (done through multiset rewriting rules without target indications) is separated from the communication (which is done through symport/antiport rules). Details can be found in Cavaliere (2003).

P AUTOMATA

The systems considered up to now are generative devices, similar to grammars: starting from an initial configuration (membranes and multisets of objects), we get sequences of transitions, hence, computations. Because of the nondeterminism, we have branching possibilities, which is why we can associate with a system a set of numbers or a set of strings — a *language*. In computability, dual to grammars we have automata, devices which recognize or analyse strings. A similar strategy has been followed also in membrane computing by introducing automata-like P systems (Csuhaj-Varju & Vaszil, 2003). In this strategy, one considers a P system of a given type (membranes, rules, multisets of objects), one inputs a given multiset w in a specified region, and if the computation ever stops, then one says that w is accepted.

The accepting behaviour is still more natural in the case of symport/antiport systems (Freund & Oswald, 2003). Just take a symport/antiport P system and consider the sequence of symbols it brings inside from the environment during a halting computation; this sequence is said to be the string recognized by the computation (if several objects are taken at the same time, then any permutation of them is allowed).

Several types of P automata were considered: of the types we have discussed, with the request to introduce the string to be analyzed symbol by symbol, at the beginning of the computation, with one-way communication among membranes, and with states associated with regions. A variant, closer to the way a problem is solved, by introducing a code of it in the initial configuration of a system, is to have a system, to introduce in the skin region a number in the form of the multiplicity of a specified object (e.g., we introduce *n* copies of an object *a*), and let the system work. If the computation stops, then the number is accepted or recognized; otherwise, the number is rejected.

Of particular interest are the systems that work in a deterministic way, where at each step one transition is possible, at most. Such systems are needed when solving problems, such as decidability problems, where we cannot accept branching, which may lead to endless computations because of "wrong" choices of rules to apply but not because the problem has no solution.

Automata-like P systems (working deterministically) are of interest also in the framework of looking for ways to "compute the uncomputable", of devising computing models able to compute beyond Turing (for instance, solving the halting problem for Turing machines, which is a problem known to be undecidable for them). The main idea used in Calude and Păun (2003) to this aim has a biological inspiration: because most reaction rates depend on the number of collisions of reactants in a time unit, the smaller the compartment, the higher the number of collisions. This means that with faster reactions, we may assume that in smaller regions (lower in a membrane structure) the time is also faster. In this way, we are led to consider P systems with different clocks in different levels of the membrane structure. If the membrane structure can grow during a computation, by membrane creation, then we can get a sufficient "time acceleration" for computing noncomputable Turing functions.

TISSUE-LIKE P SYSTEMS

The cells are in most cases living together in complex organizations — in tissues, organs, and organisms — establishing a complex communication net among them. For instance, when two protein channels from two adjacent cells come into contact (and this is enhanced by the fluid-mosaic structure or behaviour of the membranes), the two proteins often establish a common channel, by which a direct communication among the two cells can be made. Having such a channel enhances the realization of further channels, and thus a network of direct channels appears, with a specific functionality in intercellular communication (see details in Loewenstein, 1999). Rather interestingly, these channels are closed when a harmful chemical is present in a cell, and they are reopened when the "poison" vanishes. A rather similar situation appears if we take into account the organization of neurons in nets, with cells (neurons) establishing direct communication links among them through synapses, with the restriction now that we no longer have the possibility of communication through the environment (one cell expels some objects and, in the next time unit, another cell can take it from the environment. It is also natural to suppose that the communication in a neural-like net is made in a one-way manner.

These observations directly lead to considering a class of P systems which also have a natural mathematical motivation. Instead of placing the membranes in a hierarchical manner, hence in the nodes of a tree, we place them in the nodes of an arbitrary graph.

Actually, by making use of symport/antiport rules for direct communication and for communication with the environment, the communication graph is dynamically defined, depending on the rules used during a computation. Specifically, the rules used for communicating among two cells with labels i and j should specify the targets; hence, a symport rule transporting the objects of a multiset x from i to j has the form (i,x,j). If x is moved from i to j in exchange of objects of y, which are moved from j to i (this corresponds to an antiport rule), then we have a rule of the form $(i,x/y,j)$. In all cases, i and j should be different labels. One of i and j can also be equal to 0, identifying the environment.

Thus, a tissue-like P system is given by specifying the alphabet of objects, the list of cells, the sets of intercell communication rules, and the objects present initially in the environment. For each cell we have to specify the multiset of objects present in the initial configuration in the cell, as well as the rules for communication with the environment (because the targets are specified in the rules, all rules can be given as a global set for the whole system). The functioning of a tissue-like P system is again governed by the nondeterministic maximally parallel use of rules, with the result of a computation only obtained in a halting configuration. As for cell-like P systems, we can use these devices as generative mechanisms or as recognising mechanisms.

To give the reader an idea of the architecture of a tissue-like P system, we recall in Figure 5 a system from Calude and Păun (2003); it is a system able to simulate a Minsky register machine. A given number of registers are available, each one able to store a natural number. The contents of the registers are handled by a program consisting of labelled instructions, which can increase or conditionally decrease a register by one. The initial contents of a specified register are accepted if the computation halts. This system is thus capable of universal computation: the system starts with a number n introduced in cell e, and it stops if and only if the corresponding Minsky register machine stops. Here we skip the technical details, but the reader interested in mathematical developments should note that many universality results as those mentioned in a section below are proved by simulating register machines; this is always the case for automata-like P systems (and recently it was shown that most of the

Figure 5. A (deterministic, recognising) tissue-like P system

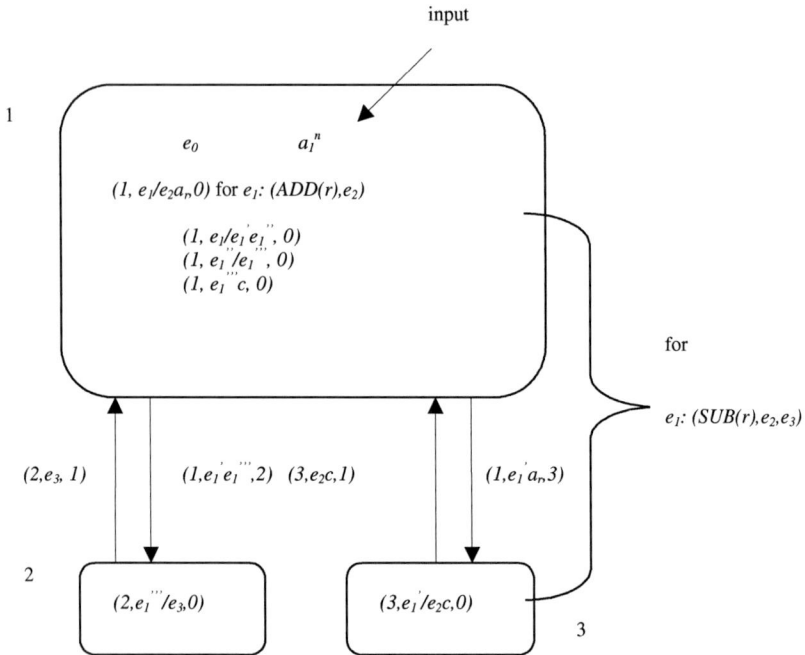

proofs in this area can be based on deterministic systems — in particular, the system from Figure 5 is deterministic).

We do not introduce here neural-like P systems, which actually do not have a well-established definition. For instance, Păun (2002) includes a chapter devoted to such systems, with states associated with neurons and with the synapses pre-established. The evolution rules are like rewriting and controlled by the states, and the communication is specified by commands *go* (meaning "go along any of the available synapses, maybe along all of them, after replication") and *out* (a way to send a result into the environment). This is rather different from the symport/antiport framework as used in the tissue-like P systems, but such a variant was briefly used in Calude and Păun (2003).

In any case, up to now, most of the research efforts were paid to the cell-like P systems (the first introduced, with many mathematical problems still not settled, and with promising applications at the level of the cell biology), while the tissue-like and the neural-like systems were considered only in a few papers, in spite of the fact that they promise to be rather useful for applications. They are not only much more flexible than the cell-like systems (we can consider cell-like systems in the nodes of an arbitrary graph, which leads to a direct extension of cell-like systems, which are then a very particular case of a tissue-like system), but they also correspond to very important real networks, such as the neural one or the Internet, which still need new mathematical models (modelling techniques). We are fairly confident that the study of tissue- and neural-like P systems both deserve further efforts and that this study will pay off both mathematically and for applications.

STRUCTURING THE OBJECTS: P SYSTEMS WITH STRING OBJECTS

In a cell, many objects can be considered "atomic" (with no internal structure), but many other objects, such as DNA molecules, have a structure, which is sometimes described by a string. This leads us to consider P systems where objects are strings — hence the evolution rules are based on string processing operations (rewriting, splicing, insertion, deletion, cut-and-paste, etc.).

For instance, rewriting rules are of the form $(X \rightarrow v; tar)$, where $X \rightarrow v$ is a usual context-free rule and *tar* is a target indication, one of *here, out,* and *in*, specifying in the standard way the region where the result of rewriting should go. We can also append to v the symbols δ and τ, which control the membrane thickness, or we can consider a priority relation among rules.

A computationally powerful idea is to combine the rewriting of strings with their duplication, considering rules of the form $r: a \rightarrow (u_1, tar_1)\|(u_2, tar_2)$. By applying r to a string $w = w'aw''$ we obtain the strings $w'u_1w''$ and $w'u_2w''$, which are sent to regions as indicated by the targets tar_1 and tar_2, respectively.

An attractive variant is to process the string objects by splicing. The formal counterpart of the recombination operation takes place for DNA molecules when they are cut by restriction enzymes and the fragments are pasted back by ligases (see Head 1987, for the initial definition of the splicing operation, and Păun et al., 1998, for a monographic presentation). This means that we have to consider rules of the form $(r; tar_1, tar_2)$, where $r = u_1\#u_2\$u_3\#u_4$ is a splicing rule (in short, $u_1\#u_2$ and $u_3\#u_4$ represent the sites where two restriction enzymes cut DNA molecules; the fragments obtained after cutting are recombined so that two possibly new strings are obtained), and tar_1 and tar_2 belong to *{here, out ,in}*; these targets specify the regions where the strings resulting from a splicing by this rule should be placed in the next configuration of the system.

In the case of P systems with string objects, the result of a computation can consist of all strings that are sent out of the system at any time during the computation (making it no longer necessary to work with halting computations). Or, when we take the number of strings into consideration — that is, we work with multisets of strings — the result is the number of strings sent out during the computation. In the latter case, it is necessary to use string-processing operations which change the number of strings. Rewriting and splicing does not have this property, but replication and splitting (cutting a string into two strings, with local changes at the cutting place) can increase the number of strings. By sending strings out of the system or storing them in certain "garbage" membranes, we can also decrease the number of string objects.

An interesting topic is processing the string objects in a parallel manner, a case in which three levels of parallelism appear: at the level of strings (several rules are applied at the same time to each string), and of regions (all strings from a region which can evolve should do it), of the whole system (all regions work simultaneously). A difficulty appears, however, with the communication, in the case when the used rules have conflicting target commands. Several solutions to this problem have been explored (see, e.g., Besozzi et al., 2003; Besozzi et al., 2004, and the bibliography therein).

The next natural step is to pass to more complex objects, in which have been discussed in several papers regarding P systems with tree and graph objects, with 2-D arrays, etc.

COMPUTATIONAL COMPLETENESS: UNIVERSALITY

As we have mentioned before, many classes of P systems, combining various ingredients, are able of simulating Turing machines; hence, they are *computationally complete*. Note that when we deal with P systems that compute numbers, we consider Turing machines as number recognizers; in the case of string objects, we can obtain the family of languages that are recognized by Turing machines (the recursively enumerable languages). Always, the proofs of results of this type are constructive, and thus have an important consequence on computability — there are *universal* (hence, programmable) P systems. In short, starting from a universal Turing machine (or an equivalent universal type-0 Chomsky grammar), we get an equivalent universal P system. This implies that in the case of Turing complete classes of P systems, the hierarchy on the number of membranes always collapses (at most at the level of the universal P systems). Actually, the number of membranes sufficient to characterize the power of Turing machines by means of P systems is always rather small; in most cases, three or four membranes suffice (in several cases, only one membrane suffices). In a few cases the best known result is six, seven, or eight membranes, but it is an open question whether or not these results are optimal. Rather interestingly, there are, however, subuniversal classes of P systems for which the number of membranes induces an infinite hierarchy of the computed sets of numbers (see Ibarra, 2004).

We only mention here, informally, some of the most interesting universality results:

1.   P systems with symbol-objects with catalytic rules, using only two catalysts, are computationally universal. The result, surprisingly strong, is the last one in a series of improvements in the number of catalysts (actually, starting with the conjecture that the catalytic systems are not Turing-equivalent).
2.   P systems with symport/antiport rules of a rather restricted size (examples: no symport, but antiport rules of weight at most 2; symport rules of weight 3 and no antiport) are universal. One membrane suffices in the cases mentioned earlier, while systems with four membranes and symport rules of weight 2 (without antiport rules) are also universal.
3.   Recently, it was proved that uniport rules and antiport rules of weight one — hence, as restricted as in biology — are universal, in systems with five membranes (this is the best result know in this moment, but it is not known to be optimal).
4.   Several types of P automata are universal, in most cases with symport/antiport rules of small weights and a reduced number of membranes. In many cases, deterministic systems were also shown to be universal.

We can conclude that the compartmental computation in a cell-like membrane structure (using various ways of communicating among compartments) is rather powerful. The "computing cell" is, effectively, a powerful "computer".

COMPUTATIONAL EFFICIENCY

The computational power (the "competence") is only one of the important questions to be dealt with when defining a new computing model. The other fundamental question concerns the computing efficiency.

A deterministic P system with catalysts and priorities, and also controlling the permeability of membranes (hence, working with symbol objects and using all basic features), can be simulated by a deterministic Turing machine with a polynomial slowdown (the "Milan Theorem", by Zandron et al., 2000). This means that by using such systems we cannot solve exponential problems in polynomial time, in spite of the fact that, exponentially, many objects can be produced in linear time by such rules as those of the form $a \rightarrow aa$. Therefore, in order to improve the computational performance of our systems, it is necessary to provide more efficient ways for producing an exponential space. Three such ways have been considered so far in the literature, and all of them were proven to lead to polynomial solutions of NP-complete problems.

These three ideas are *membrane division*, *membrane creation*, and *string replication*.

Very briefly, in the case of membrane division one uses rules of the form $[_i a]_i \rightarrow [_i b]_i [_i c]_i$. The membrane with label i is divided, and the contents of the former membrane are replicated in the two resulting membranes, with the exception of object a, which is replaced by b and c in the resulting membranes, respectively. In the case of membrane creation one uses rules of the form $a \rightarrow [_i b]_i$ (a new membrane, with label i, is created from object a), while in the case of string duplication one uses rules of the form $a \rightarrow u_1 || u_2$ (from a string xay one passes to the strings xu_1y and xu_2y, with possible targets associated with the resulting strings).

Note that both in the case of membrane division and of membrane creation several membranes may have the same label, but, because the label precisely identifies the set of rules associated with the membrane, no difficulty appears in the way the computations are defined.

By using such operations, one can obtain an exponential workspace (in the form of membranes or string objects) in a linear time, and in this way one can devise "P algorithms" which can solve NP-complete problems in polynomial (often, linear) time.

This assertion was illustrated by SAT, the Hamiltonian path problem, the node covering problem, the problem of inverting one-way functions, the subset -sum problem, and the knapsack problem. Note that the last two are numerical problems, where the answer is not "yes or no", as in decidability problems and others. Details can be found in Păun (2002) and Pérez-Jiménez et al. (2002), as well as in all the collective volumes from the bibliography.

Roughly speaking, the framework for dealing with complexity matters is that of recognizing P systems with input: a family of P systems of a given type is constructed starting from a given problem, and an instance of the problem is introduced as an input in such systems, working in a deterministic mode (or a confluent mode — some nondeterminism is allowed, provided that the branching converges after a while to a unique configuration). In a given time one of the answers, yes or no, is obtained in the form of specific objects sent to the environment. The family of systems should be constructed in a uniform mode (starting from the size of instances) by a Turing machine, working in polynomial time. A more relaxed framework is that where a *semi-uniform* construction is allowed — done in polynomial time by a Turing machine, but starting from the instance to be solved. The condition of having a polynomial time construction ensures the "honesty" of the construction: the solution to the problem cannot be found during the construction phase.

This direction of research is very active at the present moment. More and more problems are considered, the membrane computing complexity classes are refined, characterizations of the $P \neq NP$ conjecture were obtained, and improvements are sought (for instance, attempts to remove the polarizations from P systems with membrane division). Recently, a further idea to have an exponential working space was proposed: to assume that an arbitrarily large membrane system is given "for free", with all but a precise number of regions empty and to activate these regions by moving objects to them. Polynomial solutions to SAT are also obtained in this framework.

Another important recent result concerns the fact that PSPACE was shown to be included in PMC_D, the family of problems that can be solved in polynomial time by P systems with the possibility of dividing both elementary and nonelementary membranes. The PSPACE-complete problem used in this proof was QSAT (SAT with quantifiers), and the interesting conjecture was formulated that the division of nonelementary membranes are necessary to reach PSPACE (see Sosik, 2003, for details).

APPLICATIONS

Membrane computing was initiated with the goal of finding ideas, models, and tools of interest for computer science in the cell structure and functioning (and not of modelling the real cell). At the theoretical level, the goal is reached. But in the last time a double tendency is observed in the field: more and more attempts to give consistency to the applications to computer science, and more and more applications in biology.

The applications to biology are natural, if we take into account the experience of other areas, as discussed at the beginning of this chapter. Abstracting from the cell biochemistry, a new framework (starting with a new language, set of concepts, ideas, and tools) was developed that proves now to be useful for modelling not only biological processes, but also linguistic facts, management aspects, etc. Several recent applications in addressing computer science problems were reported, for instance, in sorting and ranking problems, handling 2-D languages and in computer graphics.

In many of these applications, what is actually used is the *language* of membrane computing. This means not only the long list of concepts either newly introduced or related in a new manner in this area, but also the way to represent a cell-like structure, as proposed in membrane computing. In order to illustrate these points, let us first have a partial list of concepts used in membrane computing (most of them were introduced earlier, others are self-explanatory; in order not to make the text clumsy, we do not give here further explanations and references):

membrane, region, hierarchy of membranes, skin membrane, elementary membrane, object, multiset, evolution rule, catalyst, cooperative/noncooperative rule, communication, nondeterminism, k-determinism, parallelism, configuration, transition, halting, internal output, external output, symport, antiport, uniport, carrier, promoter, inhibitor, dissolving/dividing/creating membranes, immediate communication, traveller, trace, permeability, priority, polarization, boundary rule, proton-pumping rule, gemmation, deadlock, replication, activation, merge/separate operations, confluence, cell-like/tissue-like/neural-like P system, P automaton, P transducer, one-way communication, concentration, synchronization, bistable catalyst, inter-region, channel, valuation.

In what concerns the representation possibilities, they are rather attractive for biologists—Euler-Venn diagrams, with labels for membranes, with multisets of objects (chemicals) placed in regions, and with sets of rules placed either in regions (the case of rewriting-like rules) or near membranes, to suggest that they are associated with the membranes (the case of symport/antiport rules).

However, this level of usefulness is only a preliminary one, corresponding also to the fact that the whole subject area is rather young. The next level is to use tools, techniques, and results of membrane computing. Here there appears an important question: But to what aim? Solving problems already stated by biologists, in other terms and in another framework, could be an impressive achievement, and this is the most natural way to proceed—but not necessarily the most efficient one, at least at the beginning. New tools can suggest new problems, which either cannot be formulated in a previous framework (in plain language, as is the case in biology, however specialized the specific jargon is, or using differential equations) or have no chance to be solved in the previous framework. Problems of the first type (already examined by biologists, mainly experimentally) concern, for instance, correlations of processes, of the presence/absence of certain chemicals, and their multiplicity (concentration, population) in a given compartment, while of the second type are topics related to the trajectories of biosystems when modelled as dynamical systems (e.g., a sequence of configurations can be finite or infinite, while in the latter case it can be periodic, ultimately periodic, almost periodic, quasiperiodic, etc.).

Up to now, the applications in biology follow in most cases a scenario of the following type: one examines a piece of reality, in general from the biochemistry of the cell, one writes a P system modelling the respective process, one writes a program simulating that system (or one uses one of the existing programs), and one performs a large number of experiments with the program, tuning certain parameters, and looking for the evolution of the system (usually, for the population of certain objects). Respiration in bacteria (Ardelean & Cavaliere, 2003), photosynthesis (Nishida, 2002), processes related to the immune system (Franco & Manca, 2004; Suzuki et al., 2003), and other processes (Ciobanu et al., 2003; Suzuki et al., 2001) were studied in this way. We do not recall any detail here, but we refer to the papers just mentioned, to the chapter of Păun (2002) devoted to biological applications, as well as to the papers available in the Web page devoted to membrane computing.

In any case, the investigations are somewhat preliminary, but the progresses are obvious and the hope is to have in the near future applications of an increased interest for biologists. This hope is supported also by the fact that more and more powerful simulations of various classes of P systems are available, with better and better interfaces, which allow for the friendly interaction with the program. We avoid to plainly say that we have "implementations" of P systems because of the inherent nondeterminism and the massive parallelism of the basic model — features that cannot be implemented on the usual electronic computer, but can be implemented on a dedicated, reconfigurable hardware, as done by Petreska and Teuscher (2004), or on a local network, as reported in Ciobanu and Guo (2004) and Syropoulos et al. (2004). This does not mean that simulations of P systems on usual computers are not useful; actually, such programs were used in all biological applications mentioned earlier, and can also have important didactic and research applications (see, for example, Cordón-Franco et al., 2004).

CONCLUDING REMARKS

This chapter was intended as a quick and general introduction to membrane computing, an invitation to this recent branch of natural computing, especially for the nonmathematician reader.

The starting motivation of the area was to learn from the cell biology new ideas, models, and paradigms useful for computer science — and we have informally presented a series of details of this type. The mathematical development was quite rapid, mainly with two types of results as the purpose: computational universality and computational efficiency. Recently, the domain started to be used as a framework for modelling processes from biology (but also from linguistics, management, etc.), and this is rather important in view of the fact that the P systems are reductionistic, but flexible, easily scalable, algorithmic, and intuitive models of the whole cell, while modelling the whole cell was often advocated to be an important challenge for biocomputing in the near future.

We have mentioned only a few classes of P systems, only a few types of results, and only a few of the applications reported in the literature. A detailed presentation of the whole domain is not only beyond the scope of this chapter, but also beyond the dimensions of a monograph; furthermore, the domain is fast emerging so that the reader interested in any research direction, a more theoretical or a more practical one, is advised to follow the developments through the Web page mentioned in the first section.

The presentation we have discussed was optimistic; we have only seldom mentioned weak features of membrane computing — especially from the point of view of applications in biology. Such features (the excessive reductionism, the maximality of the parallelism, the existence of the universal clock, the necessity to use noncrisp mathematics, such as probabilities, fuzzy set theory, rough set theory, the need of considering a mixture of discrete and continuous mathematics) were mentioned in several papers (part of them also at the end of the monograph of Păun, 2002). Several attempts were already made to answer these needs, and current research efforts are focused on these directions, but the reported results are preliminary, so these research areas can still be considered as open.

In short, membrane computing is a well-established branch of natural computing (of computer science in general), where plenty of things still remain to be done, and which started to prove its usefulness as a modelling framework for biology (and for other areas, too).

ACKNOWLEDGMENTS

Thanks are due to two anonymous referees for a series of useful local suggestions.

REFERENCES

Alberts, B., Johnson, A., Lewis, J., Raff, M., Roberts, K., & Walter, P. (2002). *Molecular biology of the cell* (4th ed.). New York: Garland Science.

Alhazov, A., Martín-Vide, C., & Păun, Gh., eds. (2003). Pre-proceedings of workshop on membrane computing. *WMC 2003*, Tarragona, Spain, July 2003, Technical Report 28/03, Rovira i Virgili University, Tarragona.

Ardelean, I.I., & Cavaliere, M. (2003). Modelling biological processes by using a probabilistic P system software. *Natural Computing, 2*(2), 173-197.

Bel-Enguix, G. & Gramatovici, R. (2004). Parsing with active P automata. In Martín-Vide, C., Mauri, G., Păun, Gh., Rozenberg, G., & Salomaa, A. (Eds.), *Membrane computing. International Workshop, WMC 2003*, Tarragona, Spain, revised papers. *Lecture Notes in Computer Science, 2933*, 31-42.

Besozzi, D., Ardelean, I.I., & Mauri, G. (2003). The potential of P systems for modeling the activity of mechanosensitive channels in E. coli. In Alhazov, A., Martín-Vide, C., & Păun, Gh. (Eds.), *Pre-proceedings of workshop on membrane computing, WMC 2003*, Tarragona, Spain, July 2003, Technical Report 28/03, Rovira i Virgili University, Tarragona), 84-102.

Besozzi, D., Mauri, G., & Zandron, C. (2003). Parallel rewriting P systems without target conflicts. In Păun, Gh., Rozenberg, G., Salomaa, A., & Zandron, C. (Eds.), *Membrane computing. International Workshop, WMC-CdeA 2002*, Curtea de Arges, Romania, revised papers. *Lecture Notes in Computer Science, 2597*, 119-133.

Besozzi, D., Mauri, G., Vaszil, G., & Zandron, C. (2004). Collapsing hierarchies of parallel rewriting P systems without target conflicts. In Martín-Vide, C., Mauri, G., Păun, Gh., Rozenberg, G., & Salomaa, A. (Eds.), *Membrane computing. International Workshop, WMC 2003*, Tarragona, Spain, revised papers. *Lecture Notes in Computer Science, 2933*, 55-69.

Bower, J.M. & Bolouri, H. (Eds.) (2001). *Computational modeling of genetic and biochemical networks*. Cambridge, MA: A Bradford Book, The MIT Press.

Calude, C. & Păun, Gh. (2003). Bio-steps beyond Turing. *CDMTCS Research Report 226*, University of Auckland (November 2003).

Calude, C.S., Păun, Gh., Rozenberg, G., & Salomaa, A. (Eds.) (2001). Multiset processing: Mathematical, computer science, and molecular computing points of view. *Lecture Notes in Computer Science*, 2235, Berlin: Springer.

Cavaliere, M. (2003). Evolution-communication P systems. In Păun, Gh., Rozenberg, G., Salomaa, A., & Zandron, C. (Eds.), *Membrane computing. International Workshop, WMC-CdeA 2002*, Curtea de Arges, Romania, revised papers. *Lecture Notes in Computer Science*, 2597, 134-145.

Cavaliere, M., Martín-Vide, C., & Păun, Gh., eds. (2003). *Proceedings of the brainstorming week on membrane computing*, Tarragona, Spain. Technical Report 26/03, Rovira i Virgili University, Tarragona, Spain.

Ciobanu, G. & Wenyuan, G. (2004). A P system running on a cluster of computers. In Martín-Vide, C., Mauri, G., Păun, Gh., Rozenberg, G., & Salomaa, A. (Eds.), *Membrane computing. International Workshop, WMC 2003*, Tarragona, Spain, revised papers. *Lecture Notes in Computer Science*, 2933, 123-139.

Ciobanu, G., Dumitru, D., Huzum, D., Moruz, G., & Tanasã, B. (2003). Client-server P systems in modeling molecular interaction. In Păun, Gh., Rozenberg, G., Salomaa, A., & Zandron, C. (Eds.), *Membrane computing. International Workshop, WMC-CdeA 2002*, Curtea de Arges, Romania, revised papers. *Lecture Notes in Computer Science*, 2597, 203 - 218.

Conrad, M. (1998). The price of programmability. In Herken, R. (Ed.), *The universal turing machine: A half-century survey*. Hamburg: Kammerer and Unverzagt, 285-307.

Cordón-Franco, A., Gutiérrez-Naranjo, M.A., Pérez-Jiménez, M., & Sancho-Caparrini, F. (2004). Implementing in Prolog an effective cellular solution to the knapsack problem. In Martín-Vide, C., Mauri, G., Păun, Gh., Rozenberg, G., & Salomaa, A. (Eds.), *Membrane computing. International Workshop, WMC 2003*, Tarragona, Spain, revised papers. *Lecture Notes in Computer Science*, 2933, 140-152.

Csuhaj-Varju, E. & Vaszil., G. (2003). P automata or purely communicating accepting P systems. In Păun, Gh., Rozenberg, G., Salomaa, A., & Zandron, C. (Eds.), *Membrane computing. International Workshop, WMC-CdeA 2002*, Curtea de Arges, Romania, revised papers. *Lecture Notes in Computer Science*, 2597, 203-218.

Franco, G. & Manca, V. (2004). A membrane system for the leukocyte selective recruitment. In Martín-Vide, C., Mauri, G., Păun, Gh., Rozenberg, G., & Salomaa, A. (Eds.), *Membrane computing. International Work-*

shop, WMC 2003, Tarragona, Spain, revised papers. *Lecture Notes in Computer Science*, 2933, 180-189.

Freund, R. & Oswald, M. (2003). A short note on analysing P systems. *Bulletin of the EATCS*. 78, 231-236.

Gross, M. (1998). Molecular computation. In Gramss, T., Borngoldt, S., Gross, M., Mitchell, M., & Pellizzari, Th. (Eds.), *Non-standard computation*. Weinheim, New York: Wiley-VCH.

Harel, D. (2004). A grand challenge for computing: Towards full reactive modelling of a multi-cellular animal. In Păun, Gh., Rozenberg, G., & Salomaa, A. (Eds.) (2004). *Current trends in computer science, vol. I: Algorithms and complexity*. Singapore: World Scientific, 559-568.

Head, T. (1987). Formal language theory and DNA: An analysis of the generative capacity of specific recombinant behaviors. *Bulletin of Mathematical Biology*, 49, 737-759.

Holcombe, M. (2001). Computational models of cells and tissues: Machines, agents and fungal infection. *Briefings in Bioinformatics*, 2(3), 271-278.

Ibarra, O. (2004). The number of membranes matters. In Martín-Vide, C., Mauri, G., Păun, Gh., Rozenberg, G., & Salomaa, A. (Eds.), *Membrane computing. International Workshop, WMC 2003*, Tarragona, Spain, revised papers. *Lecture Notes in Computer Science*, 2933, 218-231.

Kitano, H. (2002). Computational systems biology. *Nature*, 420(14), 206-210.

Loewenstein, W.R. (1999). *The touchstone of life: Molecular information, cell communication, and the foundations of life*. New York, Oxford: Oxford University Press.

Marcus, S. (2004). Formal languages: Foundations, prehistory, sources, and applications. In Martín-Vide, C., Mitrana, V., & Păun, Gh. (Eds.), *Formal languages and applications*. Berlin: Springer.

Martín-Vide, C., Mauri, G., Păun, Gh., Rozenberg, G., & Salomaa, A. (Eds.) (2004). *Membrane computing. International Workshop, WMC 2003*, Tarragona, Spain, revised papers. *Lecture Notes in Computer Science*, 2933, Berlin: Springer.

Nishida, T.Y. (2002). Simulations of photosynthesis by a K-subset transforming system with membranes. *Fundamenta Informaticae*, 49(1-3), 249-259.

Păun, A. & Păun, Gh. (2002). The power of communication: P systems with symport/antiport. *New Generation Computing*, 20(3), 295-306.

Păun, Gh. (2000). Computing with membranes. *Journal of Computer and System Sciences*, *61*(1), 108-143 (and Turku Center for Computer Science, TUCS Report 208, November 1998, *www.tucs.fi*).

Păun, Gh. (2002). *Computing with membranes: An introduction.* Berlin: Springer.

Păun, Gh. & Rozenberg, G. (2002). A guide to membrane computing. *Chicago Journal of Theoretical Computer Science*, *287*(1), 73-100.

Păun, Gh., Rozenberg, G., & Salomaa, A. (1998). *DNA computing: New computing paradigms.* Berlin: Springer.

Păun, Gh., Rozenberg, G., Salomaa, A., & Zandron, C. (Eds.) (2003). *Membrane computing. International Workshop, WMC-CdeA 2002*, Curtea de Arges, Romania, revised papers. *Lecture Notes in Computer Science*, 2597, Berlin: Springer.

Pérez-Jiménez, M., Romero-Jiménez, A., & Sancho-Caparrini, F. (2002). *Teoría de la Complejidad en Modelos de Computación Celular con Membranas.* Sevilla: Editorial Kronos.

Petreska, B. & Teuscher, C. (2004). A reconfigurable hardware membrane system. In Martín-Vide, C., Mauri, G., Păun, Gh., Rozenberg, G., & Salomaa, A. (Eds.), *Membrane computing. International Workshop, WMC 2003*, Tarragona, Spain, revised papers. *Lecture Notes in Computer Science*, 2933, 267-283.

Sosik, P. (2003). The computational power of cell division in P systems: Beating down parallel computers? *Natural Computing*, *2*(3), 287-298.

Suzuki, Y., Fujiwara, Y., Tanaka, H., & Takabayashi, J. (2001). Artificial life applications of a class of P systems: Abstract rewriting systems on multisets. In Calude, C.S., Păun, Gh., Rozenberg, G., & Salomaa, A. (Eds.), *Multiset processing: Mathematical, computer science, and molecular computing points of view. Lecture Notes in Computer Science*, 2235, 299-346.

Suzuki, Y., Ogishima, S., & Tanaka, H. (2003). Modeling the p53 signaling network by using P systems. In Alhazov, A., Martín-Vide, C., & Păun, Gh. (Eds.), *Pre-proceedings of workshop on membrane computing, WMC 2003,* Tarragona, Spain, July 2003, Technical Report 28/03, Rovira i Virgili University, Tarragona, 449-454.

Syropoulos, A., Allilomes, P.C., Mamatas, E.G., & Sotiriades, K.T. (2004). A distributed simulation of P systems. In Martín-Vide, C., Mauri, G., Păun, Gh., Rozenberg, G., & Salomaa, A. (Eds.), *Membrane computing. International Workshop, WMC 2003*, Tarragona, Spain, revised papers. *Lecture Notes in Computer Science*, 2933, 355-366.

Tomita, M. (2001). Whole-cell simulation: A grand challenge of the 21st century. *Trends in Biotechnology*, 19, 205-210.

Webb, K. & White, T. (2004). UML as a cell and biochemistry modeling language. *BioSystems*, to appear.

Zandron, C., Ferretti, C., & Mauri, G. (2000). Solving NP-complete problems using P systems with active membranes. In Antoniou, I., Calude, C.S., & Dinneen, M.J. (Eds.), *Unconventional models of computation*. London: Springer, 289-301.

Chapter II

State Transition Dynamics:
Basic Concepts and Molecular Computing Perspectives

Vincenzo Manca, University of Verona, Italy

Giuditta Franco, University of Verona, Italy

Giuseppe Scollo, University of Verona, Italy

ABSTRACT

Classical dynamics concepts are analysed in the basic mathematical setting of state transition systems where time and space are both completely discrete and no structure is assumed on the state's space. Interesting relationships between attractors and recurrence are identified and some features of chaos are expressed in simple, set theoretic terms. String dynamics is proposed as a unifying concept for dynamical systems arising from discrete models of computation, together with illustrative examples. The relevance of state transition systems and string dynamics is discussed from the perspective of molecular computing.

INTRODUCTION

A dynamical system is a structure that changes in time. This characterization immediately shows the wide range of such a notion. A ball that is moving, a flow of electrons in a conductor, the propagation of a wave, a chemical reaction, and even a chess game, a process running on a computer, or the development and the life of an organism are examples of dynamical systems. What is essential in these systems are three components: *space,* collecting all the possible states of the system; *time,* collecting the different instants at which the system is considered; and *dynamics*, which associates, to each instant, the corresponding state of the system. The notion of state is to be viewed in a wide sense — it could be a set of *microstates*, or a probability distribution on a set of states, or, as in the case of a quantum systems, a *superposition* of states expressed by a vector in a suitable vector space. In all cases, the general structure of such a system is given by a triple $D = (T, S, \delta)$, where the dynamics δ is a function from the set T of instants to the set S of states. The various kinds of dynamical systems are essentially determined by the structure of the space, the nature of the time, and the way dynamics is characterized.

In the classical theory of dynamical systems suggested by physics, and naturally described by differential equations, time is a continuous entity and instants are the points of the real axis. When the instants are identified with natural or integer numbers, then the systems are usually called *discrete dynamical systems*, in contrast with the former ones, which are called *continuous dynamical systems*. But, in both cases, space is always assumed to be a continuous metric space (the usual three-dimensional space, or a Hilbert phase space of possibly infinite dimension). In this paper, we address the problem of considering, in general terms, dynamical systems that are completely discrete in that, not only are the instants natural or integer numbers, but also the space is a discrete entity. We know that any discrete information can be encoded by a string over a suitable alphabet V; therefore, generally, a discrete space is identified by a subset of V^* (the set of all the strings over an alphabet V).

We are typically interested in the behaviour of dynamical systems — that is, how their configuration changes in time. When the dynamics is given explicitly, then the behaviour of the system is completely described by it, and, in more precise terms, a triple time-space-dynamics determines a *global dynamical system*. But, in almost all cases, the dynamics is given implicitly because we only have some conditions $\Phi(\delta)$ that specify it. Therefore, the main problem to solve is just the computation of the values of the dynamics in time, starting from an initial instant and using *local* conditions that implicitly determine it — in other words, the reconstruction of a global dynamics from a local one. In classical terms, where conditions $\Phi(\delta)$ are expressed by means of some derivatives of d, this corresponds to the Cauchy problem of differential equations that, under suitable hypotheses, can be shown to have a unique solution. In general terms, a *local* dynamical system is expressed by a structure $(T, S, \Phi(\delta))$, where $\Phi(\delta)$ are conditions that determine a function from T to S, or more generally, in the case of a nondeterministic system, from T to the parts of S.

PRELIMINARIES

A way to overcome the difficulties of integration in classical dynamical systems is that of adopting approximation techniques from numerical analysis and computational mathematics that provide a partial answer to the Cauchy problem by means of a discrete version of it. Therefore, a natural question arises: Why not formulate directly, in discrete terms, continuous dynamical systems that in general are difficult to integrate? Could their behaviours be determined by specific algorithms in this way? This viewpoint needs further clarification. Classical, continuous dynamical systems are often solved in an analogical way by physical simulation — that is, by mapping the states of a given system M into those of a physical model F, where some constraints are imposed to mimic those of the original system. In this way, starting from an initial condition of F that translates an initial condition of M, the system F is observed in its evolution. If at a given instant t its state $s(t)$ is determined, then its inverse translation as a state of M gives (up to measurement errors) a knowledge of M at time t. Now, if we translate the states of a system into strings over some alphabet and translate the local conditions on its dynamics by means of relations between strings, then the symbolic run that starts from an initial string is the discrete analogue of a physical simulation, let us say a *symbolic simulation*. A game that transforms strings can thus "generate" the behaviour of a given system. The problem is that of finding good models that, case by case, fit the relevant features of the system we want to symbolically simulate.

The dynamics of discrete systems, besides being instrumental to their direct algorithmic simulation, often prove to be a very natural representation of biomolecular dynamics, where the symbolic entities which come into play are easily amenable to strings or closely related structures, or to dynamical networks in some cases. Models hitherto applied with remarkable success are cellular automata (CA), Lindenmayer systems (which are a special case of CA), Kauffman networks (Kauffman, 2000), and various kinds of automata and transition networks. However, typical properties that are relevant to computational models (such as termination, confluence, and reducibility) are replaced in those systems, when viewed as dynamical systems, by other properties, that share nontermination as their outstanding prominent feature. These properties include periodicity, recurrence, emergence, propagation, stability, and evolution.

The aforementioned properties deserve deep insight and ability to interconnect them if one aims at modelling biological systems whose complexity often wholly exceeds the capabilities of continuous methods. Now, the identification of elementary forms of biological behaviour, together with analysis criteria and parameters that could enable their classification, stands out as a target that would prove greatly useful towards the understanding of which basic principles govern the complexity of the living systems (Bernardini & Manca, 2003). The aforementioned discrete models yet suffer from the lack of a basic conceptual framework. In continuous system dynamics, a few fundamental concepts have been identified that are based upon the metric and topological structure of state spaces, and precisely those concepts enabled the powerful development of so-called qualitative dynamics, first presented by Poincaré. By contrast, when speaking of an attractor, or of chaotic or complex behaviour of a cellular automaton or Kauffman network, the meaning of those terms is to be understood either vaguely or in an *ad hoc* manner, viz. one that is closely related to, thereby dependent on, the specific situation at hand. In a fundamental pioneering work (Ashby, 1956) state transition graphs, called kinetic graphs, are introduced in the context of dynamical concepts. However, despite its strong biological relevance, the book does not cope with the mathematical aspects of its conceptual apparatus. The most interesting results on dynamical indicators have been identified by Wolfram (1986) and Langton (1990), noting that they discriminate classes of behaviour which appear to be different, but with no formal definition of the specific characteristics of a given behaviour. Important experimental results in this context have been reported in Kauffman (2000) and Wuensche (1998).

This chapter undertakes a systematic investigation of dynamical concepts that is aimed at building a basic conceptual framework for discrete system dynamics.

The initial standpoint is that of making no assumptions whatsoever about the nature of system states, while trying to characterize fundamental concepts of dynamics in purely set theoretic terms. In such a context only a function is assumed to specify the transition from any given state to a set of possible states, and time proceeds along the iteration of that function. We shall see that, in spite of the minimal generality of such a basic assumption, a surprising wealth of concepts arises, whose dynamic valence enjoys both evidence and depth. These concepts are introduced in the next section, together with a preliminary analysis which we plan to further develop in future work. In the subsequent section we move from dynamics based upon, so to say, atomic states, to dynamics where, on the contrary, a discrete structure over the states is assumed. We highlight how all string manipulation systems from formal language theory can be naturally expressed as string transition systems. From this perspective, notable concepts in grammars and automata theory take up a purely dynamic character. For instance, the language generated by a grammar is an attractor of a certain kind. Examples of simple dynamic systems that naturally enjoy a biomolecular meaning conclude this section, while highlighting the interest to their treatment in purely discrete terms. Finally, we give a simple result that sheds light on a computational limitation of periodicity. We review and discuss the outcomes from our analysis in the conclusions, where perspectives for further research are briefly drawn.

STATE TRANSITION DYNAMICS

The main task of this section is to define typical dynamical concepts (orbit, basin, periodicity, attractor, initial data sensitivity) in the framework of state transition systems, where we need to cope with two basic aspects that are completely different from the classical frameworks: space is discrete, and dynamics is nondeterministic, in the sense that each state can evolve towards many possible states. To this aim, a notion of orbit that generalizes the classical concept of orbit to the nondeterministic case, will turn out to be essential.

Definition 1. A *state transition dynamics* is a pair *(S, q)* where *S* is a set of states and *q* is a function from *S* into its power set.

We call *quasi state* any subset of S. However, we will speak of states also in the case of quasi states when from the context it is clear that we refer to a set of states rather than a single state. If the dynamics function *q* is total, then we say that the dynamics is *eternal*. We speak of *q*-dynamical systems in order to mention the dynamics function *q* explicitly. We also write $x \rightarrow y$ whenever $y \in q(x)$, with *q* understood from the context; "$\rightarrow$" thus denotes the transition relation of the *q*-dynamical system. Given a quasi state *X*, then we put $q(X) =$

$\cup_{s \in X} q(s)$. As customary, q^i denotes the i-fold iterated composition of q, and we indicate $q^*(X) = \cup_{i \in N} q^i(X)$. We write $q^*(x)$ rather than $q^*(\{x\})$ for simplicity.

Definition 2. Given a q-dynamical system, an *orbit* is a sequence X_0, X_1, ... of quasi-states such that $X_{i+1} = q(X_i)$ for all $i \geq 0$. An orbit of origin s, shortly an s-orbit, is an orbit such that $X_0 = \{s\}$. $O(s)$ is the set of orbits of origin s.

Definition 3. Given a q-dynamical system, a *trajectory* of origin s, shortly a s-trajectory, is a function ξ from the set N of the natural numbers to the set S of the states, such that $\xi(0) = s$, and $\xi(i+1) \in q(\xi(i))$ for all $i \geq 0$. We call *jet* of origin s (*s-jet*) the set $J(s)$ of all trajectories of origin s.

It is clear that the notion of dynamical systems defined earlier is nondeterministic because any state can transform into a set of possible states. However, according to a standard technique in automata theory, it is easy to map a q-dynamical system to a deterministic system where states are the quasi states of the original system, and in this case any orbit of the original system turns out to be a trajectory of the new system, where orbits are equivalent to trajectories, but the nondeterministic aspect is essential for enlarging the possibilities of modelling real-world phenomena.

Definition 4. An orbit of origin X is *periodic* if, for some $n > 0$, $q^n(X) = X$. The orbit is *eventually periodic* if for some k, $n > 0$ $q^{n+k}(X) = q^k(X)$. In this case k is called the *transient* and n the *period*.

It is easy to restrict the concepts of periodicity and eventual periodicity to the special case of trajectories. Specializing or extending the periodic state's definition can lead to several special cases of periodicity.

Definition 5. Given two orbits $(X_i \mid i \in N)$ and $(X'_i \mid i \in N)$, we say that the first one is *included* in the second one if $X_i \subseteq X'_i$ for every i. The first orbit is *eventually included* in the second one if there exists a number j such that $X_k \subseteq X'_k$ for all $k \geq j$.

Of course, the property of inclusion is a particular case ($j = 0$) of the property of eventual inclusion. In the special case of periodic or eventually periodic orbits, the verification of eventual inclusion is immediate because we need to check whether the languages of the period (of finite length) are included.

In the following definition we adopt the usual notation $O(f)$ and $\Omega(f)$ (Rosen, 2000) of asymptotic algorithmic complexity, which classifies the growth of real functions with respect to a real function f from natural numbers to positive real numbers.

Definition 6. Let D be a q-dynamical system, and μ a function from the set of states of D to the real numbers, that in this context is called a *Ljapunov function* over D. An s-orbit is $O(f(n))$-*divergent* with respect to the Ljapunov function μ if $\mu(q^n(s))$ has order $O(f(n))$, and it is $\Omega(f(n))$-*divergent* with respect to μ if $\mu(q^n(s))$ has order $\Omega(f(n))$.

Definition 7. A state is said to be *final* if the transition relation is not defined in it. A state s is a *fixed point* if the transition relation transforms it into itself deterministically — that is, $q(s) = \{s\}$.

Any transition relation can be naturally extended to another one in which all the final states become fixed points; in this manner any orbit becomes infinite. In the following we assume dynamics to be eternal (i.e., without final states).

Definition 8. Let D be a q-dynamical system. A nonempty set B of states of D is a *basin* if, for every state x of B, $q(x)$ is included in B.

Definition 9. Let D be a q-dynamical system and B a basin of D.

i)   A *potential attracting set* A of B is a nonempty subset of B such that for every state x of B there is a state z in some state of the x-orbit such that the z-orbit is eventually included in A (i.e., after a finite number of steps, the state of the z-orbit is included in A, and so are all further states in it).

ii)   An *unavoidable attracting set* A of B is a nonempty subset of B such that for every state x of B, the x-orbit is eventually included in A.

The following definition supports the search for minimal cases in relation with the concepts just introduced, for a given basin.

Definition 10. Let D be a q-dynamical system, B a basin in D.

i)   We say that a subset Y of a potential attracting set A of B is $\Diamond$-*removable* (read "may-removable") *from* A w.r.t. B, if $A \backslash Y$ is a potential attracting set of B; a state $y \in A$ is $\Diamond$-removable from A w.r.t. B if so is $\{y\}$.

ii) We say that a subset Y of an unavoidable attracting set A of B, []-*removable* (read "must-removable") *from A w.r.t. B*, if $A\backslash Y$ is an unavoidable attracting set of B; a state $y \in A$ is []-removable from A w.r.t. B if so is $\{y\}$.

Note that in any of the variants of the previous definition, if Y is infinite, although it may consist of removable states, it may prove to be nonremovable from A w.r.t. B. The following example illustrates such a case. Some celestial terminology is first introduced.

Definition 11. Let D be a q-dynamical system and x a state of D. An x-trajectory ξ is an x-*flight* if it is an injective function on N. An x-*flight* ξ is an x-*blackhole* if $q^*(x) \subseteq \xi(N)$.

We simply speak about flight and blackhole when we do not want to specify the origin state.

Example 1. Let the basin B be a flight in the dynamical system D. Every state is both $\lozenge$-removable and []-removable from B w.r.t. B itself, indeed so is every finite subset of B. However, any infinite subset of B is neither $\lozenge$-removable nor []-removable from B w.r.t. B.

Definition 12. Let D be a q-dynamical system and B a basin of D.

i) A potential attracting set A of B is a *potential attractor* of B if it is minimal under set inclusion, meaning no nonempty subset of A is $\lozenge$-removable from A w.r.t. B.

ii) An unavoidable attracting set A of B is an *unavoidable attractor* of B if it is minimal under set inclusion, meaning no nonempty subset of A is []-removable from A w.r.t. B.

When we say "attracting set" and "attractor" without any attribute, then "potential" is implicitly understood. In particular, an attractor of basin B is an attracting set of B with no $\lozenge$-removable states w.r.t. B. A similar special case holds for unavoidable attractors.

In the following, we assume that all of the concepts relative to a given basin are defined in relation to a fixed basin B, when not otherwise mentioned.

Definition 13. Let D be a q-dynamical system.

i)   A state x is *recurrent* if $x \in q^n(x)$ for some $n>0$. Henceforth let $R_\Diamond$ denote the set of recurrent states in a fixed basin B of D.

ii) A state x is *eternally recurrent* if $\forall n>0: y \in q^n(x) \Rightarrow \exists m>0\ x \in q^m(y)$. Henceforth let $R_{[]}$ denote the set of eternally recurrent states in a fixed basin B of D.

A state is recurrent if and only if it occurs infinitely often in its own orbit. While this is a necessary condition for eternal recurrence in the context of eternal dynamics, it is not sufficient. For example, consider the two-state dynamical system specified by: $x{\to}x, x{\to}y, y{\to}y$; here x is a recurrent state, yet not an eternally recurrent one.

The following proposition collects several useful observations, which easily follow from the previous definitions.

Proposition 1. Let D be a q-dynamical system.

i)   Every basin is an unavoidable attracting set of itself.
ii)   Every unavoidable attracting set of basin B is an attracting set of B.
iii) If $Y \subseteq A$ is []-removable from A w.r.t. basin B, then Y is also $\Diamond$-removable from A w.r.t. B.
iv)   An attractor of basin B is unique, if it exists; we then speak of *the* attractor $A_\Diamond$ of B, while $A_\Diamond = \varnothing$ means that B has *no* attractor.
v)   An unavoidable attractor of basin B is unique, if it exists; $A_{[]}$ then denotes *the* unavoidable attractor of B, while $A_{[]} = \varnothing$ means that B has no unavoidable attractor.
vi)   The (unavoidable) attractor of a basin B is also the (unavoidable) attractor of any subset of B that is a basin and includes the (unavoidable) attractor. In particular, every (unavoidable) attractor is also its own (unavoidable) attractor.
vii) Every eternally recurrent state is recurrent, thus $R_{[]} \subseteq R_\Diamond$.

In the presence of certain kinds of flights in a basin B (see Example 1), it may happen that $A_\Diamond = \varnothing$ while $A_{[]} \neq \varnothing$, as well as that $A_\Diamond \neq \varnothing$ while $A_{[]} = \varnothing$. Examples of these phenomena are deferred until after a useful characterization of removability of states, in any of the variants of Definition 10.

Proposition 2. Let D be a q-dynamical system and B a basin of D.

i)   For any attracting set A of B, a state $y \in A$ is $\lozenge$-removable from A w.r.t. B if and only if $\forall x \in B \, \exists z \in q^*(x) \, y \notin q^*(z)$.

ii) For any unavoidable attracting set A of B, a state $y \in A$ is $[]$-removable from A w.r.t. B if and only if for no $x \in B$ y occurs infinitely often in the x-orbit.

Proof. Statement ii follows immediately from definitions 10.ii and 9.ii. As one may expect, statement i follows from definitions 10.i and 9.i, but the derivation is slightly less immediate since the displayed characteristic condition is a simpler, equivalent formulation of a somewhat lengthier condition. The latter, whose equivalence with $\lozenge$-removability of y from A w.r.t. B is easily derived from definitions 10.i and 9.i, is as follows: $\forall x \in B \, \exists z \in q^*(x)$ such that y does not occur infinitely often in the z-orbit. Now, the existence of each z with the property just stated is equivalent to require that, for some $n \geq 0$ (that for which $z \in q^n(x)$) and some $m \geq 0$ (that for which $\forall k \geq m \, y \notin q^k(z)$), there exists some $z' \in q^{m+n}(x)$ such that $y \notin q^*(z')$. The equivalence of this requirement to the simpler characteristic condition displayed in the statement is immediately apparent.

Proposition 2.i helps us to characterize $\lozenge$-nonremovability of states. The following fact will simplify the reasoning.

Proposition 3. $R_{[]} \neq q^*(R_{[]})$ for any q-dynamical system D then basin B of D.

Proof. When $R_{[]} = \varnothing$, let $x \in q^*(R_{[]})$ and either $x \in R_{[]}$ or $x \in q^n(y)$ for some $n > 0$ and eternally recurrent state y, in which case $y \in q^*(x)$. Then, for every $z \in q^*(x)$, $x \in q^n(y)$ entails $z \in q^*(y)$; hence $y \in q^*(z)$ by eternal recurrence of y, $x \in q^n(y)$ entails $x \in q^*(z)$, and x is eternally recurrent as well.

Every eternally recurrent state is $\lozenge$-nonremovable, since the characteristic condition stated by Proposition 2.i for $\lozenge$-removability fails at the state itself: if $y \in R_{[]}$, then no $z \in q^*(y)$ can ever be found such that $y \notin q^*(z)$, since y is eternally recurrent. Regardless of the presence of flights, $R_{[]}$ is always "the" set of $\lozenge$-nonremovable states, as it is easy to check by using Definition 13.ii and Proposition 2.i.

In a similar way, Proposition 2.ii helps us to characterize $[]$-nonremovability of states, but under the assumption of absence of flights; it immediately entails that recurrent states are []-nonremovable and, more generally, that no state in $q^*(R_\lozenge)$ is []-removable, since each of them occurs infinitely often in the orbit having origin at some (recurrent) state. In the absence of flights, $q^*(R_\lozenge)$ is "the" set of []-nonremovable states, while flights introduce further possibilities of []-nonremovability of states, as shown by the following case.

Example 2. Let the basin B consist of flights $X = (x_n \mid n \in N)$ and $Z = (z_n \mid n \in N)$, with the implied transitions $x_n \to x_{n+1}, z_n \to z_{n+1}$ for all $n \in N$, the transitions $x_n \to z_0$ for all $n \in N$, and no other transitions. $R_\lozenge$ is empty in this case (the transition system is cycle-free), and so are $R_{[]}$ (by Proposition 1.vii) and $q^*(R_\lozenge)$. Every state in B is $\lozenge$-removable (from B w.r.t. B itself). But while every state in X is also []-removable, no state in Z is []-removable. Furthermore, every subset of X is $\lozenge$-removable, but only the finite ones are []-removable, which entails that B has no unavoidable attractor. Finally, only the finite subsets of Z are $\lozenge$-removable, so B also has no attractor.

As stated earlier, in the presence of flights it may happen that the unavoidable attractor does not exist while the attractor does, as well as that the latter does not exist while the former does. Here are examples of these phenomena.

Example 3. Let the basin B consist of a flight $X = (x_n \mid n \in N)$ and a fixed point z such that $x_n \to z$ for all $n \in N$. It is easy to check that $R_{[]} = R_\lozenge = \{z\}$, and $A_{[]} = \varnothing$ while $A_\lozenge = \{z\}$. The unavoidable attractor does not exist because every state in the flight X is []-removable, while no infinite subset of any flight is ever []-removable.

Example 4. Let the basin B consist of two different connected components: a fixed point z and a flight $X = (x_n \mid n \in N)$ where only x_0 is recurrent, leaving $x_0 \to x_0$. Then $R_{[]} = \{z\}$, $R_\lozenge = \{x_0, z\}$, and $A_{[]} = B$ while $A_\lozenge = \varnothing$. In this case the unavoidable attractor exists, by the recurrence of x_0 that makes every state in the flight []-nonremovable ($q^*(R_\lozenge) = B$ in this case), whereas the attractor does not exist, since every state in the flight X is $\lozenge$-removable, while no infinite subset of it is $\lozenge$-removable in this case.

The cases we have illustrated naturally call for a more precise characterization of the existence of attractors, since they show that the presence of a flight may, but need not, hamper the validity of $A_{[]} = q^*(R_\lozenge)$ and of $A_\lozenge = R_{[]}$. To satisfy this curiosity, the following concepts prove purposeful.

Definition 14. Let B be a basin, and X a flight $(x_n \mid n \in N)$ in B.

i) X is *recurrent in B* if $x_n \in q^*(R_\diamond)$ for some $n \geq 0$.
ii) X is *eternally recurrent in* B if $x_0 \in R_{[]}$ (hence $x_n \in R_{[]}$ for all $n \in N$).

The desired characterization is provided by the following fact, which somehow mimics the famous Poincaré's recurrence theorem (Devaney, 1989).

Proposition 4. Let D be a q-dynamical system and B be a basin of it.

i) $A_{[]} = q^*(R_\diamond)$ if every flight in B is recurrent, otherwise $A_{[]} = \emptyset$.
ii) $A_\diamond = R_{[]}$ if every blackhole in B is eternally recurrent, otherwise $A_\diamond = \emptyset$.

The previous proposition does not only characterize the *existence* of (unavoidable) attractors, but also their *extent* as (closures of) sets of recurrent states. Periodic orbits and fixed points are simple cases of attractors, according to the following proposition.

Proposition 5. Any eventually periodic orbit determines an attractor, consisting of the states of the period, where the basin comprises the states of the transient and of the period.

Proof. Eventual periodicity rules out the existence of flights.

Definition 15. A *fixed point attractor* is an attractor that only consists of fixed points.

Periods, fixed points, and eternally recurrent blackholes give rise to particular cases of attractors, under an appropriate choice of the basin.

Definition 16. Two orbits of origins s_1 and s_2 are *intersecting* if there are m, $k \geq 0$ such that $q^m(s_1) \cap q^k(s_2) \neq \emptyset$.

Summing up the results given so far we obtain the following classification of attractors:

Proposition 6. A q-dynamical system may have three different types of attractors:

1.   Periodic attractors — that is, periodic orbits (fixed point attractors are a special case thereof).
2.   Eternally recurrent blackholes.
3.   Complex attractors, or a combination of the two previous cases, not necessarily disjointed ones.

Attractors of the third type arise from the closure property of attractors. For example, combining periodic orbits or an eternally recurrent blackhole with periodic orbits gives rise to complex forms of attractors. Generally, they may comprise the composition of many periods according to different structures, and in extreme cases they present a special case of chaotic complexity where the behavior becomes completely unpredictable.

In a next section we try to outline some features of discrete chaos where our nondeterministic notion of orbit seems to clarify a deep informational aspect of this phenomenon.

CHAOS IN Q DYNAMICS

The notion of chaos is one of the most important concepts in the theory of dynamical systems. The famous Lorenz "strange" attractor (Devaney, 1989) was the first clear identification of a phenomenon, of a meteorological interest, where a deterministic law provides a behaviour that is intrinsically unpredictable. In fact, although the system, defined on a set of continuous states, is governed by simple deterministic rules, its evolution cannot be calculated because even a small inaccuracy in the determination of its initial state (that is unavoidable for the finiteness of state representations) will be amplified in such a way that the future state will have no correlation with the initial one.

Dynamical systems reveal many aspects of chaotic behaviour, but a univocal formal definition of chaos is so far a matter of speculation and of mathematical research. Moreover, the classical tradition of dynamical systems has mainly developed mathematical theories of chaos for continuous systems. Therefore, how do we formalize a notion of chaos at a general level that can include chaotic aspects of both time- and space-discrete systems? In several papers, when a discrete dynamical system is referred to as chaotic, this attribute is only used with an informal meaning as synonymous to "unpredictable" and "erratic". More specifically, three requirements seem to be essential in a chaotic dynamics, as they are expressed in classical definitions for continuous systems: a chaotic system on a space X is topologically transitive (there exists an orbit that is dense in it), the system has sensitivity to initial conditions, and the set of its periodic points is dense in X (Bonanno & Manca, 2002; Devaney, 1989).

1. **Global recurrence.** In a chaotic dynamical system the set of all states is its own attractor, which is also called a *strange* attractor. In other words, a chaotic behaviour is a global property that cannot be decomposed into distinct parts. It follows that, according to the analysis developed in the previous section, in a chaotic dynamical system all states are eternally recurrent states.

2. **Sensitivity to initial conditions.** This requirement implies an exponential divergence of orbits where points that are "near" become exponentially far in time. This aspect can be viewed as an explosion of orbits, or as an "informational drift" along the orbits, and it is relative to some Ljapunov function with respect to some measure of the closeness of states.

For example, a random variant of the shift map of symbolic dynamics (Devaney, 1989), restricted to a set of finite strings on $\{0,1\}$, deletes the first symbol at a fixed end of the given string and adds a random bit at the other end of the string. In a dynamics based on the shift map, a bit of information is thus lost at each step; hence all the information that identifies a state vanishes in a number n of steps equal to the length of the string — that is, all 2^n strings of length n collapse to the same state.

3. **Ubiquitous periodicity.** This property refers to the erratic aspect of chaos: orbits are wandering everywhere and forever, so explosions of orbits are mixed with orbit implosions in such a way that dynamics returns periodically onto itself, according to their intrinsic recurrence. But these periods are endlessly overlapping each other.

Now we try to express these requirements in the context of state transition dynamics.

Definition 17. A dynamical system D of states S is *dynamically transitive* if there is a point a such that $q^*(a) = S$.

Definition 18. A dynamical system D is *μ-sensitive*, with respect to a Ljapunov function μ, if all orbits of D are exponentially divergent with respect to μ.

The following is a notion related to sensitivity that can be expressed in set theoretic terms.

Definition 19. A dynamical system D of states S is *erratic* if for any two states s_1 and $s_2 \in S$ there is a positive number n such that $q^n(s_1)$ and $q^n(s_2)$ are disjoint sets. The smallest n that satisfies this condition is the *divergence time* $d(s_1, s_2)$ of the two points.

The function d measures the "explosion" of a system and can be compared in several ways, such as in terms of growth order, with other parameters of the system, in order to characterize the level of erraticism of the system.

Definition 20. A dynamical system D of states S is *ubiquitously periodic* if any two of its periodic orbits eventually intersect.

Chaos, or several forms of it, can be defined by combining the features defined earlier according to different values of their intrinsic parameters (that specify different degrees of these features).

Informally, when chaos appears, there is no way to characterize any set of states according to any temporal property of theirs, nor conversely, any set of instants of time according to any spatial property. In other words, space is lost in time and time is lost in space, in such a way that no clock driven by observation of the system can predict its behaviour, and no clock can be extracted from the system that can be used in the observation of other dynamics. In particular, no kind of ordering relation can be defined on instants of time and no time arrow is definable by means of the system. It is interesting to note that, in Greek cosmogony, Chaos precedes the birth of Kronos.

However, the given notion of chaos can be better analysed by giving some chaos indexes: 1) an index for measuring the extent to which orbits diverge with respect to the considered Ljapunov function, and 2) an index for measuring the average intersection between periodic orbits, or the extent to which periodic orbits overlap, thereby confusing any localization of the period while the system is running through it at each time.

It is interesting to realize that the definition of chaos expressed in terms of state transition dynamics allows us to consider chaos of dynamical systems very similarly to the notion of deterministic chaos of continuous systems such as logistic maps, Bernoulli shifts, and Manneville maps (Devaney, 1989), which, although ruled by very simple dynamics, present evident chaotic behaviours. What makes these systems intrinsically chaotic is the essential role of quasi states in their descriptions. In fact, their dynamical systems are defined on states given by real numbers, but these numbers are always expressed by some of their finite approximations, or rational numbers. In this sense, one may view a set of states as a rational number — a quasi-state — that comprises all the real numbers that, at some level of approximation, share the same rational number. Therefore, the sensitivity to initial conditions corresponds to the exponential growth of a Ljapunov function along the orbits associated to the finite approximations of the states of the system. Analogously, the overlapping of periods in these systems corresponds to the eventual intersections of their periodic orbits, when we consider the quasi-states that correspond to the finite representations of their states.

In conclusion, what is called deterministic chaos does not differ from nondeterministic chaos. The difference is only a matter of the way orbits are defined. In deterministic chaos these are introduced by the intrinsic approximation of states; in the nondeterministic chaos of state transition dynamics, the orbits are defined by state transition relations that provide many possible states that can be reached from a single state. But what is very important to remark is that determinism is not synonymous with predictability, nor is nondeterminism synonymous with unpredictability. Indeed, a system that is deterministic but chaotic becomes unpredictable, and a nondeterministic system can be predictable in several aspects.

STRING TRANSITION SYSTEMS

In the following section, we assume all basic notions and notations of formal language theory, computability and computational complexity (see Rozenberg & Salomaa, 1997; Rosen, 2000, for more details). In particular we use the usual regular operations on languages: concatenation (indicated by juxtaposition), union (indicated by $+$), finite iteration (indicated by $*$), and infinite iteration (indicated by ω) (van Leeuwen, 1990). When we assume states to be expressed by strings, then the natural way to define a local completely discrete dynamical system D is that of identifying it with a *string transition system*, specified by a set S of strings (states of the system) over an alphabet V and a binary transition relation $\rightarrow$ on S:

$$D = (V, S, \rightarrow)$$

A Cauchy system is obtained by adding an initial string s_0 to it:

$$D = (V, S, s_0, \rightarrow)$$

In the following, we assume dynamical systems and related Cauchy systems in this restricted sense. State transition dynamics in the sense of Definition 1 is easily recognized in these structures, by taking the image of each state under the transition relation as its image under the dynamics function q.

String dynamical systems can be conveniently described by *transition graphs* in several manners. The simplest form of a transition graph is given by a set of nodes that represent the states of the system and a set of arcs connecting them. An arc from a node to another node represents the transition from a state to another state. For instance, the following two examples present transition graphs, given by a list of arcs between pairs of nodes that express very simple dynamics. The arrow before s_0 indicates that s_0 is the initial state.

Example 5. (*2-Cycle*) $\rightarrow s_0 \rightarrow s_1, \quad s_1 \rightarrow s_0$

Example 6. (*3-Cycle*) $\rightarrow s_0 \rightarrow s_1, s_1 \rightarrow s_2, s_2 \rightarrow s_0$

When the set of states is infinite, the previous representation can no longer be used. In this case, however, states can often be identified as paths in a finite transition graph. Namely, strings are associated to arc labels and the concatenation of the labels along a path represents the state identified by the path. This technique works whenever the set of states can be represented by a set of strings generated by a Chomsky grammar. An example of this possibility is the following graph, where arc labels are displayed as arrow superscripts, and the empty string is assumed where no superscript occurs. An arrow after a target node, such as $s_2\rightarrow$, indicates that the node is an *output node*, meaning that the concatenation of labels of a path from the initial node to an output node is a *generated string*. In the following notation, the graph expresses a Büchi automaton (van Leeuwen, 1990) where every path from the initial node to the output node can be extended to an infinite path where the output node occurs infinitely often.

Example 7. $\rightarrow s_0 \rightarrow^a s_0, s_0 \rightarrow^b s_0, s_0 \rightarrow^a s_1, s_1 \rightarrow^b s_2 \rightarrow, s_2 \rightarrow s_0$

It is easy to check that the set of generated strings is described by the regular expression $((a+b)^*ab)^*$, and the infinite paths of the graph, starting from s_0, are described by the extended regular expression $((a+b)^*ab)^\omega$. Let us extend this example by adding a numeric label to the first two rules, to specify an upper bound on the number of applications of each rule, say 5 in both cases:

$$\rightarrow s_0 \rightarrow^a_5 s_0, s_0 \rightarrow^b_5 s_0.$$

In this case, the infinite paths of the graph generate infinite strings that could be described by the following expression:

$$((a+b)^{*(a \leq 5,\, b \leq 5)}\, ab)^{\omega}$$

where a finite iteration may apply to $(a+b)$ subject to the condition that it makes neither the number of occurrences of a nor that of b become greater than 5. This behaviour has a very interesting property, called "almost periodicity" (Marcus & Păun, 1994). In fact, we are sure that between two occurrences of ab there is a distance (number of symbols) of at most 10 elements.

An important aspect of the dynamics described by the graph of Example 7 is its strong nondeterminism, in that not only along a path we are free to choose among different arcs exiting from a given node, but moreover the same symbol can be generated by using different arcs. We show that in this case we can avoid the second kind of nondeterminism, by taking a graph where different arcs exiting from any given node generate different symbols, but the price we pay is a more complex graph:

$$\rightarrow s_0 \xrightarrow{a} s_1,\, s_0 \xrightarrow{b} s_0,\, s_1 \xrightarrow{a} s_1,\, s_1 \xrightarrow{b} s_2 \rightarrow,\, s_2 \xrightarrow{a} s_1,\, s_2 \xrightarrow{b} s_0$$

This example shows that the increasing of nondeterminism can obtain the same behaviour, but in a simpler way.

Another way to describe transition graphs is that of associating strings to nodes and conditions to the arcs that say how to get strings in the transition along that arc. In this way, states are the strings associated to the last node of a path in the graph. In the following example a transition graph is given, where states are the strings obtained from the initial symbol S of a grammar by applying its rules $S := aSb$, $S := ab$.

Example 8. $\rightarrow s_0 \xrightarrow{S} s_1,\, s_1 \xrightarrow{S:\, =\, aSb} s_2,\, s_1 \xrightarrow{S:\, =\, ab} s_2,\, s_2 \rightarrow s_1,\, s_2 \xrightarrow{T^*} s_3 \rightarrow$

Superscripts of arcs with the symbol := specify replacement rules that can be applied to strings associated to the source node in order to get strings associated to the target node. We associate the empty string λ to the initial state s_0 and with the first transition rule a symbol S is associated to node s_1. In the fourth rule no change is performed on the strings associated to node s_2 and they are by default associated to the target node s_1; while in the last transition rule the condition T^* says that some strings associated to node s_2 can become associated to s_3 only if all their symbols belong to the set $T = \{a,b\}$ of terminal symbols of the grammar. In this manner, along the paths from s_0 to s_3 we get as strings associated to s_3 all the strings generated by the grammar.

The language generated by a Chomsky grammar can be defined as the fixed point attractor of the start string σ (that is, of the basin $q^*(\sigma)$), if we extend the transition relation of the grammar by turning the strings of terminal symbols into fixed points. Similarly, we could represent string dynamical systems associated to many types of automata, transducers, matrix grammars, L-systems (Rozenberg & Salomaa, 1997, and Rosen, 2000), P-systems (Păun, 2002), cellular automata (Wolfram, 1986), X-machines (Gheorghe, 2000), Petri Nets (Reisig, 1985), Kauffman Nets (Kauffman, 2000), PB systems (Bernardini & Manca, 2002), and PBE systems (Bernardini & Manca, 2003) where an input-output interaction with the environment is allowed during the evolution.

In the following example we propose a system that has interesting natural interpretations — a chemical one and a biological one, since it can be seen as a model of an unstable catalytic transformation or of the diffusion of an infection.

Example 9. Assuming that initially positive quantities of both symbols C and G are present, the following rules may apply:

$$CG \rightarrow GG$$
$$C \rightarrow C$$
$$G \rightarrow \lambda$$
$$G \rightarrow K$$
$$G \rightarrow G$$

When agent G is present, then agent C transforms into G; this is the behaviour of a prion that transforms another protein into prion, by a sort of contagion, or the communication of an infection from a sick individual (symbol G) to a healthy one (symbol C). Agent G is unstable until two possible transformations — it either dies or survives; in the latter case, it remains immune to further contagion and mutates into an allosteric form of C. This is recognized in our system in that, nondeterministically, G may either disappear or transform into symbol K, which represents a healthy and immune individual.

If a state is the pair of numbers of healthy individuals and sick ones, that is (|C|+|K|, |G|), then we obtain the underlying system of linear differential equations:

$$x' = axy - bx$$
$$y' = -axy$$

where x is the number of sick individuals and y is the number of healthy nonimmune individuals. This system, where x_0 and y_0 are the values for $t=0$ and integration is understood from 0 to t, is not integrable, in the sense that it has no explicit form:

$$x(t) = x_0 \, e^{\int (ay(s) - b) \, ds}$$
$$y(t) = y_0 \, e^{\int -ax(s) \, ds}$$

But from qualitative analysis we know that, for some values of the parameters a and b, and in particular when a/b is large enough, the process of infection ends in finite time with a positive quantity of healthy and nonimmune individuals.

It will be an aim of future work of ours to interpret and analyse the dynamics of this system by means of the concepts proposed in this paper, whereby we expect to gain more information.

PERIODICITY AND UNIVERSALITY

Periodicity and eventual periodicity are properties with a strong computational significance; we will show that, in a suitable computational framework, the periodicity decision problem turns out to be computationally equivalent to the termination problem. Let us outline the basic idea underlying this result.

A computational formalism, considered as a set C of machines is said to be *computationally universal* if to any machine M of C a language $L(M)$ can be associated in such a way that the class RE of recursively enumerable languages is given by:

$$RE = \{ L(M) \mid M \in C\}$$

Let us assume that machines of C have the following property (typical of usual computational formalisms): to any $M \in C$ a finite but unbounded "work space" $w(M)$ is available, sufficient to hold any finite string on some finite alphabet that the operation of the machine may require. This workspace is further split into two unbounded parts, sharing a common boundary point, in such a way that there exists another machine $M' \in C$ that can simulate the behaviour of M by using only one of the two unbounded parts of $w(M')$, while leaving the other one free.

Now, let us consider a machine M of a computational formalism C, and let us consider the dynamical system $D(M)$ naturally associated to M, where machine states are the states of the machine M used during its work, starting from an initial state s_0. We show that the general problem of deciding whether $D(M)$ is eventually periodic can be translated into the problem of deciding if, for a suitable machine M' of C, the computation starting from s_0 is terminating or not.

Proposition 7. If C is a computationally universal class of machines, then the eventual periodicity of the related dynamical systems is not decidable.

Proof outline. Given a machine H of C we can always construct the following machine H' that behaves exactly as machine H does, but by simulating H on just one of the two unbounded parts of its work space $w(H)$. On the remaining part of $w(H)$, a counter is implemented that at each step increases its value. When H reaches a final configuration — that is, a configuration where the computation of H stops — H' differs from H. according to the following strategy: if q_f is the final state of H, machine H' stops updating the counter and starts oscillating between q_f and a new state. The existence of the counter guarantees that H' is eventually periodic if and only if H reaches state q_f.

Now let us suppose that we have a decision procedure P for the eventual periodicity of the dynamical systems $\{D(M)|M \in C\}$, such that $P(D(M)) = 1$ if $D(M)$ is eventually periodic and $P(D(M)) = 0$ otherwise. This would imply that we can decide the termination for any machine M of C. In fact, in order to decide whether a machine M will terminate, it is sufficient to know the value of $P(D(M'))$. But the computational universality of C contradicts the existence of such a possibility.

In conclusion, if a class C of machines is computationally universal, then the eventual periodicity of the associated dynamical systems is not decidable.

It is easy to change the previous proof in order to show the following proposition as well.

Proposition 8. If C is a computationally universal class of machines, then the periodicity of the related dynamical systems is not decidable.

CONCLUSION

The various types of attractors that are offsprings of our analysis are forms of behaviour that a system inevitably uncovers in its temporal unfolding. The various kinds of recurrence and dynamic connections between space regions also yield central concepts for dynamic characterization. A key point of this analysis was the notion of the quasi-state. This analysis yields a rich scenario, including other concepts that are crucial in biological dynamics, such as bifurcation, intermittency, and dissipation.

String transition systems are a specialization of state transition dynamics and can be applied to formal language theory, but we envisage a wider applicability of these systems. A fundamental aim of dynamical analysis is the definition of parameters for classification of behaviours that are relevant to specific finalities. Both cellular automata and Kauffman networks explain that the relationship between the transition function and the state structure strongly determines dynamically relevant qualities. We put forward that several parameters that are identified in those contexts, such as locality, arity, channelling, majority, and input entropy (Kauffman, 2000; Wuensche, 2002), could be easily generalized and analysed in terms of string transition systems. We plan to develop this intuition in later publications.

Finally, the dynamics of biological adaptive systems can be viewed as a computation where the "result", which a system searches for, is not a state but a dynamical pattern — or better, a stable pattern — that is an attractor that fulfills some "enjoyable" conditions. Computability theory and dynamical systems have subtle, at times twisted, links with certain informational features that are in the scope of both (Casti, 1995). After all, two great unpredictability results from the past century that have definitely uncovered the limitations of the Laplacian view of science are the computability limits discovered by Turing and the deterministic chaos by Lorenz. Perhaps both stem from a common root, which comes across inherent limits of information (Dufort & Lumsden, 1997). The fact that periodicity is as undecidable as termination seems to be a perfectly natural truth after all; but bringing it to the fore may foster reflection and insight into deeper, analogous phenomena.

REFERENCES

Ashby, W. R. (1956). *An introduction to cybernetics*. London: Chapman and Hall.

Bernardini, F. & Manca, V. (2002). P systems with boundary rules. In Păun, Gh., Rozenberg, G., Salomaa, A., & Zandron, C. (Eds.), *Membrane computing. International Workshop, WMC-CdeA 2002*, Curtea de Arges, Romania, revised papers. *Lecture Notes in Computer Science*, 2597, 107-118.

Bernardini, F. & Manca, V. (2003). Dynamical aspects of P systems. *BioSystems, 70* (2), 85-93.

Bonanno, C. & Manca, V. (2002). Discrete dynamics in biological models. *Romanian Journal of Information Science and Technology*, (5) 1-2, 45-67.

Casti, J. L. (1995). *Complexification*. New York: HarperPerennial.

Devaney, R. L. (1989). *Introduction to chaotic dynamical systems*. Boston: Addison-Wesley.

Dufort, P. A. & Lumsden, C. J. (1997). Dynamics, complexity and computation. In Lumsden, C. J., Bandts, W. A., & Trainor, L. H. (Eds.), *Physical theory in biology*. River Edge, NJ: World Scientific.

Gheorghe, M. (2000). Generalized stream X-machines and cooperating distributed grammar systems. *Formal Aspects of Computing*, 12, 459-472.

Kauffman, S. (2000). *Investigations*. Oxford: Oxford University Press.

Langton, C. G. (1990). Computation at the edge of chaos: Phase transitions and emergent computation. *Physica D*, 42, 12-37.

Marcus, S. & Păun, Gh. (1994). Infinite (almost periodic) words, formal languages and dynamical systems. *Bulletin of the EATCS*, 54, 224-231.

Păun, Gh. (2002). *Membrane computing: An introduction*. Berlin: Springer.

Reisig, W. (1985). *Petri nets: An introduction*. EATCS monograph on theoretical computer science. Berlin: Springer.

Rosen, K. H. (2000). *Handbook of discrete and combinatorial mathematics*. Boca Raton, FL: CRC Press.

Rozenberg, G., & Salomaa, A. (1997). *Formal language theory, vol. 1–3*. Berlin: Springer.

van Leeuwen, J. (1990). *Handbook of theoretical computer science, vol B: Formal models and semantics*. Cambridge, MA: MIT Press.

Wolfram, S. (1986). *Theory and application of cellular automata*. Boston: Addison-Wesley.

Wuensche, A. (1998). Discrete dynamical networks and their attractor basins. *Complexity International* (online journal).

Wuensche, A. (2002). Basins of attraction in network dynamics: A conceptual framework for biomolecular networks. In Schlosser, G. & Wagner, G.P. (Eds.), *Modularity in development and evolution*. Chicago: University of Chicago Press.

Chapter III

DNA Computing and Errors:
A Computer Science Perspective

Lila Kari, The University of Western Ontario, Canada

Elena Losseva, The University of Western Ontario, Canada

Petr Sosík,
Silesian University, Czech Republic and
The University of Western Ontario, Canada

ABSTRACT

This chapter looks at the question of managing errors that arise in DNA-based computation. Due to the inaccuracy of biochemical reactions, the experimental implementation of a DNA computation may lead to incorrectly calculated results. This chapter explores different methods that can assist in the reduction of such occurrences. The solutions to the problem of erroneous biocomputations are presented from the perspective of computer science techniques. Three main aspects of dealing with errors are covered: software simulations, algorithmic approaches, and theoretical methods. The objective of this survey is to explain how these tools can reduce errors associated with DNA computing.

INTRODUCTION

Biomolecular computing is a field that studies biologically based computational paradigms that serve as alternatives to the traditional electronic ones. Biomolecular computing includes DNA computing (Adleman, 1994; Head, 1987), RNA computing (Faulhammer et al., 2000), peptide computing (Balan et al., 2002), and membrane computing (Păun, 2000). The main idea behind DNA computing is that data can be encoded in DNA strands and molecular biology tools can be used to perform arithmetic and logic operations.

Nearly a decade has passed since the field of DNA computing premiered on the scientific stage as the possible computational paradigm of the future. The idea attracted research from a wide spectrum of mathematical and natural sciences. However, the inherently complex nature of biological processes tempered the advancement of the field, suggesting the development of biocomputing will trail a path that is different from that of electronic computing half a century ago. It is becoming increasingly more apparent that most plausible implementations of biocomputing are likely to produce some unexpected and erroneous results. From chemical reactions in vitro that occasionally have unpredicted output, to unforeseen problems in vivo, it seems that many errors are not only inevitable but also an integral part of the biological processes. The purpose of this chapter is to provide a survey of the tools that computer scientists offer for dealing with the imminent problem of managing errors in DNA computing.

The battle for reliability of biomolecular computation and reduction of errors can be fought on several fronts. Research is conducted to find better ways to encode information in DNA, to develop more efficient algorithms, and to improve laboratory techniques, among other results. This survey does not cover the wide scope of research in chemistry, biology, physics, or engineering that contributes to dealing with errors in biomolecular computing. Instead, this exposition explores the tools that computer science offers us in managing the errors that arise in DNA computing processes.

A single strand of DNA (deoxyribonucleic acid) is a molecule made of a sequence of nucleotides, also called bases. Four types of nucleotides are present in DNA, called adenine, guanine, cytosine, and thymine. These are abbreviated as *A, G, C,* and *T* respectively. A single strand of DNA is held together by covalent bonds that keep the bases linearly attached to each other. In addition, it is possible for hydrogen bonds to form between the *A* and *T* bases, as well as between *C* and *G* bases of two different strands. This property is referred to as the *complementarity* of nucleotides — that is, *A* and *T* are said to be complementary, and so are the *C* and *G* bases. Bonds between complementary nucleotides are called *base-pair* bonds. Unlike beads on a string, a sequence of nucleotides is distinct from its reverse. This property of a DNA strand is called the *polarity* of a strand, and it imposes a distinction on the two ends of the DNA molecule. The two ends of the strand are called the *3'-end* and the *5'-end*. Whenever the nucleotides of two sequences are complementary and the strands have opposite polarities (orientation in space), the strands will anneal (hybridize) to form a double helix. Sequences with this property are also called "Watson-Crick complementary" in honour of the two scientists who discovered the structure of DNA. See Watson et al. (1987) for further information on molecular biology.

Hybridization is one of the fundamental mechanisms used in DNA-based computing. Using hybridization, along with other biochemical operations, potentially general-purpose computations can be carried out (Freund et al., 1999). However, many of the designed experiments fail to produce the anticipated computational answer. In this chapter, the discussion of errors in DNA computing carries a computational connotation and refers to events that lead to obtaining a computationally incorrect result. The source of errors leading to incorrect answers can be anything from an inappropriate choice of encodings of information into DNA strands to unsuitable experimental conditions. Here we examine reducing the effect of such errors on the correctness of the computed answer.

This chapter addresses the various aspects of managing errors in three main sections examining software simulation, algorithmic, and theoretical approaches. The software simulation tools described in the first section can accomplish such activities as testing computation protocols. This testing verifies protocol correctness before it is carried out in a laboratory experiment, thus detecting potential errors. Certain errors in DNA computation can be avoided by designing strands that prevent the formation of DNA secondary structures (intramolecular bonds). Algorithms for constructing DNA sequences with this property are mentioned in the second section of the chapter. Finally, the last section gives an overview of theoretical methods aimed at reducing errors caused by undesirable hybridization. This includes template-based sequence design of code words and a study of bond-free DNA languages, followed by a discussion of future trends and research directions in the area.

More precisely, the software section of this paper looks at three different programs: BIND, SCAN, and Edna. This is not a complete list of programs written for DNA computing purposes, but it provides an overview of the types of problems that can be successfully addressed with software. The BIND program's main focus is to estimate DNA hybridization temperatures. Hybridization is a reaction present in all DNA computing protocols. Understanding under what conditions hybridization occurs involves knowing the hybridization temperature of each reaction. Carrying out laboratory experiments at inappropriate temperatures is a common source of erroneous results. Prediction of hybridization temperatures by BIND helps to avoid such problems.

A short overview of thermodynamics of DNA hybridization is also included in this section. The SCAN program is designed to find DNA sequences for computation that meet a required set of constraints. These constraints include, for example, the property of strands that avoid formation of secondary structures. Another constraint is that computation rules, encoded in DNA sequences, should not interfere with each other. If these constraints are not met, the computation is likely to result in errors. Finally, the simulation software Edna can test DNA-based algorithms for possible errors. Edna simulates biochemical processes and reactions that can occur during a laboratory experiment. Testing laboratory protocols with Edna before the experimental implementation is conducted can avoid many errors.

Another avenue of research aimed at reducing errors in DNA computing looks at methods of reducing the possibility secondary structures of DNA strands. In particular, this problem arises when a number of short DNA strands attach together to form long strands. While the original strands may not sick to each other in undesirable ways, the resulting strand may form bulges or loops. The structure freeness problem for combinatorial sets asks whether, given a set of DNA words, a concatenation of an arbitrary number of words from this set will form a word that leads to secondary structures. Algorithms that answer this question are based on heuristic calculations of the free energy of a DNA strand.

An entirely different approach to reducing errors related to DNA computing is offered by theoretical computer science methods. The question of developing appropriate techniques for encoding data in DNA can be studied in both the formal language theory and the coding theory frameworks. The final section of the chapter first explains a template-based design of DNA sequences, followed by an overview of the properties of DNA languages.

SOFTWARE APPROACHES

BIND Simulator

Hybridization of DNA strands is utilized in virtually all proposals for DNA computation—both experimental and theoretical ones. Hybridization, otherwise called annealing, is the process that forms a double-stranded DNA helix from two single-stranded DNA sequences, provided that certain conditions apply. Almost all models of computing with DNA rely on an accurate prediction of whether some DNA sequences will anneal. The success of a given model, therefore, directly depends on the correctness of this prediction.

While it is not easy to determine whether two arbitrary sequences will anneal, some general principles can be considered. We can see if two single strands of DNA anneal by checking if they are Watson-Crick complementary. However, the picture of DNA hybridization is much more complex than that. The length of the sequences makes a difference. If one sequence is longer than another and the double strand has an unhybridized segment on its end—called the sticky end—the stability of the helix is also affected. The concentration of strands in the solution, the temperature at which the reaction takes place, and numerous other factors also play a role. To complicate the situation further, it is also possible for nucleotides to form bonds with nucleotides other than their complements. This situation is called a *base-pairing mismatch*. To get the full picture, new software tools are needed to predict the likelihood of hybridization of two strands.

One example of such software is BIND (Hartemink & Gifford, 1997), which uses the Nearest Neighbour Model (NNM) of annealing to describe hybridization of strands. For two DNA sequences, we can say that there exists a temperature at which half of the strands in the solution are hybridized and half are not. This temperature is called the *melting temperature* for the given DNA double helix. Melting is the opposite process of hybridization; it separates a double strand into two single strands. The melting temperature is denoted by T_M. NNM investigates thermodynamics of DNA hybridization and provides a method for calculating T_M. We shall now explain the principles of this model and how it is used by the BIND software to determine T_M.

Throughout this paper we will use a convention to write $n_1n_2...n_k/m_1m_2...m_k$ to denote the DNA duplex (i.e., a double-helical segment) formed by two complementary strands $5'- n_1n_2...n_k - 3'$ and $3'- m_1m_2...m_k - 5'$, where n_i and m_j are individual nucleotides. When the duplex is formed with a self-complementary strand (i.e., $n_1n_2...n_k = m_km_{k-1}...m_1$), the duplex is written simply as $n_1n_2...n_k$.

Originally introduced by Borer et al. (1974), NNM proposes that the most significant contribution to helix stability comes from the order of nucleotides in the helix. Helices with exactly the same base-pair composition can have sufficiently different melting temperatures (SantaLucia et al., 1996). The difference in melting temperatures is attributed to the ordering of base pairs. The term *stacking interactions* is used in reference to the processes affecting the stability of base-pair bonds as a result of neighbouring base-pairs interactions. The order in which the base pairs are stacked is a primary factor influencing duplex stability, according to NNM.

The basic idea of the model is that for short, single-stranded complementary sequences, hybridization happens like the closing of a zipper. Base-pair bonds form one by one, gradually closing the DNA "zipper". The model views hybridization of two single strands as a sequence of smaller subreactions, each one corresponding to the formation of a single base-pair bond. With this in mind, the model uses characteristics of the smaller subreactions to estimate melting temperature and other properties of the entire hybridization reaction.

To calculate the melting temperature for a strand, BIND considers the thermodynamics of DNA hybridization. A single reaction of base-pair formation is the formation of hydrogen bonds between two complementary bases. Each reaction has a number of characteristics associated with it, including enthalpy, entropy, and Gibbs free energy. Thermodynamics is a study of the interconversions of various types of energy, and since these characteristics deal with changes in the energy of the system, they are called the *thermodynamic parameters* of the model.

We now explain these thermodynamic parameters. Enthalpy change, denoted by $\Delta H°$, of a reaction is the amount of heat released (exothermic reaction) or absorbed (endothermic reaction) by the system. Entropy is a measure of randomness or disorder. Spontaneous changes can be accompanied by either an increase or a decrease of entropy in the system. Change in entropy is denoted by $\Delta S°$, and hybridization of strands is a process increasing the order of the system; therefore, $\Delta S°$ of hybridization reactions has a negative value (corresponding to a decrease in disorder). Gibbs free energy (or simply, free energy) describes the potential of a reaction to occur spontaneously. Each chemical reaction that converts products into reactants also happens in the reverse direction simultaneously, but at a different rate. When the rates are equal, the system is in equilibrium.

A simple example of a system in equilibrium is a bucket of water at zero degrees with some ice in it. While nothing happens visibly, there are two reactions going on — some ice is melting and some water is freezing. DNA hybridization works in a similar manner. The melting temperature of a DNA helix is defined as the temperature at which half of the DNA strands in the solution are annealed and half are not. The system is at equilibrium. Some single strands are annealing, some double strands are melting, but the rates of reaction of both forward (annealing) and reverse (melting) reactions are the same. When a system is not at equilibrium, one direction of the reaction is spontaneous; it is the one we observe. The change in Gibbs free energy, denoted by $\Delta G°$, determines whether the system is at equilibrium or not. $\Delta G°$ is the free energy of products minus the free energy of reactants. In our case, the products are the annealed double strands and the reactants are the single double strands. For the forward direction to occur spontaneously, $\Delta G°$ has to be negative. When $\Delta G°$ is zero, the system is at equilibrium. If $\Delta G°$ is positive, then the reverse reaction (melting) will occur spontaneously.

Gibbs free energy has a close correlation to melting temperature. The stronger the bond of a DNA duplex (double-stranded segment), the higher its melting temperature and the greater the change in free energy of the hybridization reaction. Another way to say this is that a duplex folds into a structure that has the lowest free energy.

So how does the BIND software use these thermodynamic parameters to calculate melting temperature? As already mentioned, NNM used by BIND views hybridization as the closing of a zipper. The formation of each new base-pair bond is examined as a separate reaction. The complete sequence of base-pair formation reactions together makes up the hybridization process. For example, hybridization of *ATG/TAC* is viewed as the initiation reaction forming *A/T*, then forming *T/A*, and finally *G/C*. The three mini-reactions together are equivalent to the hybridization reaction. In this model, the base-pair bond formation reaction depends only on the base pair formed immediately prior to it, but not on any of the future ones. That is, the free ends of the zipper do not participate in the formation of a single base-pair bond and neither does the other, closed end of the zipper. The only factor in a new base-pair bond formation reaction is the preceding base pair next to which the new pair is stacked. That immediately preceding base pair is called the *nearest neighbour* and is the source of the model's name.

$\Delta G°$ for the hybridization of the entire duplex is calculated as the sum of $\Delta G°$ of the mini-reactions plus some extra parameters accounting for initiation of the first pair. The other two parameters are calculated similarly. BIND calculates the melting temperature T_M as follows:

$$T_M = \frac{\Delta H°}{\Delta S° + R \ln([C_T]/4)}$$

In this equation, R is the Boltzmann's constant and C_T is strand concentration. If [Na⁺] concentrations differ from $1M$, there is an adjustment term, which is added to the equation.

The BIND program can help test any computational protocol that is based on site-specific annealing. For example, it has been used to verify a sticker-based model of DNA computation described by Roweis et al. (1996). The model design involved a long template sequence and five short DNA sequences that are supposed to anneal to the template at specific locations. These locations are exactly complementary to the sticker sequences. BIND successfully verified that the sticker sequences would not be able to incorrectly anneal at any other template location. The program predicted melting temperatures for the sticker sequences and produced a temperature range at which no erroneous binding could occur.

The SCAN Program

One of the common problems with designing DNA-based computational components is to select those sequences that are best suited for computation and yield the most reliable results. The SCAN program (Hartemink et al., 1999) assists in this task by scanning a vast space of possible designs and selecting those that meet a large set of constraints.

Consider the design of a unary counter, as proposed by Hartemink et al. (1999). The design is based on the technique of programmed mutagenesis, which is a systematic rewriting of a DNA sequence based on a set of rules. The rewriting is sequential, with rules devised in a way that allows all rules to be present in the solution. At a given step in the computation, only the valid rules can enter into a reaction. The unary counter consists of a sequence, called a template, of 12mer DNA strands. (A 12mer is a single strand of 12 nucleotides.) There are three types of participating 12mers, denoted X, Y, and Z, with Z representing number zero and X and Y representing number one. The initial template contains only the Z 12mers. At each consequent stage of the computation, one Z 12mer is replaced by either X or Y, incrementing the counter by one. This particular design employs two rules, encoded as sequences called *primers*. Each step of the computation involves the annealing of one of the primers to the current template and an extension of this primer to synthesize a new strand. After the DNA duplex is melted, this extended strand becomes the new template in the subsequent step of computation.

In this design, the first decision that has to be made is to select the X, Y, and Z sequences, as well as the primers. The fundamental property of programmed mutagenesis is that annealing of a primer to the template needs to include mismatches; otherwise, the original template would never be modified. Programmed mutagenesis is an example of a method that uses errors in biomolecular operations to facilitate the essential features of the computation.

The selection of a mismatch location plays an important role. If the bond between the primer and the template is too unstable, the extension of the primer will not be successful. The SCAN program considers many candidate sequences for the unary counter and selects the best ones. Another condition that must be met by the design is that the sequences used as rules of computation must be present in the solution simultaneously. The rules that are not actively used in a computational step should not be interfering. In particular, inactive rules should not be able to bind to the template sequence at any location. The sequences used in the design must have a low chance of forming secondary structures. This is necessary since a sequence that binds to itself at any point in the computation becomes unusable. Finally, this design of the unary counter is included as part of a plasmid, or a circular DNA molecule. That is, the DNA sequence encoding the unary counter is incorporated into the sequence of the plasmid. This plasmid needs to be chosen so that primers do not bind to it. The SCAN program tests all of these constraints. For the unary counter, SCAN examined over 7.5 billion design candidates and narrowed the choice down to nine candidates that satisfied all of the required constraints.

Edna

One of the most natural attempts to understand and improve the processes involved in DNA computing is to simulate the procedures and chemical reactions that take place in the laboratory. An actual laboratory experiment, whether successful or not, may take weeks or even months to conduct. At this stage of biomolecular computing, advancements in the field are made through trial and error processes. A large number of trials to learn from is therefore conducive to further progress. Having simulation software that can closely resemble laboratory procedures allows the opportunity to gain insight into wet lab experiments without spending the time required to carry out the experiments themselves.

One example of such software is Edna (Garzon & Oehmen, 2001), a simulation tool that uses a cluster of PCs and demonstrates the processes that could happen in test tubes. Edna can be used to determine if a particular choice of encoding strategy is appropriate, to test a proposed protocol and estimate its performance and reliability, and even to help assess the complexity of the protocols. Test tube operations are assigned a cost that takes into account many of the reaction conditions. The measure of complexity used by Edna is the sum of these costs added up over all operations in a protocol. Other features offered by the software allow the prediction of DNA melting temperature, taking into account various reaction conditions. One of the crucial properties of Edna is that all molecular interactions are local and reflect the randomness inherent in biomolecular processes. The test tube reactions simulated by Edna can be carried out in the virtual test tubes under a variety of reaction conditions. Temperature, salt concentrations, and strand concentrations can be adjusted as necessary. Testing the scalability of a proposed protocol is another application of Edna.

In addition to the software described (BIND, SCAN, and Edna), a number of other software packages can aid in biomolecular computing. A DNA sequence compiler (Feldkamp et al., 2001a), a DNA sequence generator (Feldkamp et al., 2001b), and the NACST/Seq sequence design system (Kim et al., 2002) are some examples of these.

ALGORITHMIC METHODS

Structure-Free DNA Word Sets

It is possible that within a single-stranded DNA molecule some segments will bind to other segments of the same molecule, forming loops, bulges, and other shapes. This process involves the formation of base pairs within a single DNA strand and the folded structure is referred to as the *secondary structure* of DNA (see Figure 1). For a given molecule, there may be several possible secondary structures. Moreover, each formed structure is not necessarily stable; it may partially fold, then unfold, and, finally, refold into a different structure. Exactly how this happens depends on which segments of the molecule are Watson-Crick complementary and where these segments are located within the molecule.

Formation of secondary structures is an important factor to be considered in the designs of DNA-based computation. Suppose a strand is intended to be used for computation by interacting with other strands, and instead it folds into a secondary structure. It follows that this strand becomes useless for computation and introduces errors into the process. Therefore, a major effort in DNA computing is directed towards the study of how to predict and avoid secondary structures in DNA.

While there have been algorithms proposed for predicting the secondary structure of RNA and DNA, a new twist to the problem arises in the context of DNA computing. The structure freeness problem is proposed by Andronescu et al. (2003). Suppose that we have a set S_i containing DNA single strands, and they all have length l_i. We have t such sets — that is $1 \le i \le t$. Note that the lengths l_i need not be the same. Let S be the cross product of these sets, or

Figure 1. Secondary structure of a DNA molecule (dotted lines are base pairs, and solid lines form the backbone)

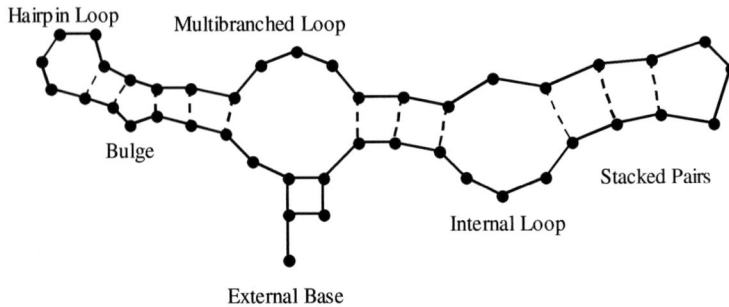

$S = S_1 \times S_2 \times ... \times S_t$. The structure freeness problem for combinatorial sets asks if the strands in the set S are free from secondary structure. The goal is to choose sets S_i in a way that ensures no secondary structure for strands from S.

One example where the structure freeness problem arose in DNA computation was in solving the satisfiability problem (Braich et al., 2001). The satisfiability problem asks, given Boolean expression, if there exist value assignments that make this formula true. The solution involved six word pairs, where each word is a 15mer. These words were used to construct new strands by choosing one word from each pair and concatenating these six words to form long strands of length 6 * 15 nucleotides. Since there are two words to choose from and six pairs in total, there are 2^6 resultant strands. This is a particular instance of the structure freeness problem as formulated earlier, and it can be solved using the algorithm by Andronescu et al. (2003).

The algorithm is based on calculating the free energy of the secondary structure and uses thermodynamic parameters somewhat similarly to the techniques used in predicting the melting temperature of a DNA duplex, as described earlier. DNA tends to fold into structures with the lowest free energy. The free energy of a loop can be experimentally established, which is done for common types of loops, such as hairpin loops, multibranched loops, internal loops, bulges, and other shapes encountered in secondary structures. The free energy of the secondary structure is calculated as a sum of the free energies of the component loops. In this model it is assumed that the free energy of a given loop does not depend on other loops. The strand is said to have secondary structure if the free energy corresponding to that strand is positive, and is said to be structure-free if the free energy is a negative value.

The algorithm for solving the structure freeness problem is based on dynamic programming techniques. It runs in $O(n^3)$ time and is a practical tool for testing if a proposed word set meets the required condition of avoiding secondary structures in computation.

Design of DNA Code Words

In surface-based strategies for DNA computing, DNA molecules are attached to a surface, such as a chemically modified gold film (Liu et al., 1998). A large quantity of unattached DNA molecules is placed in the same solution. Some of these molecules can be selected via hybridization to the complementary surface molecules and a subsequent removal of unhybridized molecules. Selected molecules in this case represent solutions to combinatorial problems.

In these types of experiments there is an assumption that a strand will bind to its perfect Watson-Crick complement. This is not always the case and can be the source of potential errors. Suppose there are molecules in the solution that differ by only one nucleotide. It is possible that the surface molecule complementary to one of them can just as easily bind to another molecule. The problem is that a mismatch in a single nucleotide location does not prevent hybridization from occurring. One way to avoid this situation is to ensure that any two molecules in the solution differ in more than one location. This property can be formalized in terms of the Hamming distance, denoted $H(w_1, w_2)$, and defined as the number of locations in which two given words w_1 and w_2 are distinct. In this definition, the two words must be of equal length. $H(w_1, w_2) \geq d$ means that word w_1 differs from word w_2 in at least d places. For a set of words S, the Hamming distance constraint (HD) requires that any two words w_1 and w_2 in the set S have $H(w_1, w_2) \geq d$.

The other type of error that can arise in surface-based computing is for the unattached molecules in the solution to bind to each other, instead of hybridizing to those on the surface. This occurs between molecules w_1 and w_2 that are Watson-Crick complementary. More specifically, w_1 is the reverse complement of w_2, which is defined as the reversed sequence of w_2, where each nucleotide is replaced by its complement. The reverse complement of w_2 is denoted by $wcc(w_2)$. The reverse complement Hamming distance constraint (RC) requires that for any two words w_1 and w_2 in the set S, we have $H(w_1, wcc(w_2)) \geq d$.

Another consideration in surface-based computing is that hybridization to different surface strands should occur simultaneously. This implies that respective melting temperatures should be comparable for all hybridization reactions that are taking place. This is the third main constraint that the set of words under consideration needs to adhere to.

To address the design of DNA code words according to these three constraints, Tulpan et al. (2003) propose an algorithm based on a stochastic local search method. The melting temperature constraint is simplified to the constraint requiring that the number of C and G nucleotides is 50 percent. The algorithm produces a set of DNA words that satisfy one, two, or all three of the noted constraints, as required. The difficulty with nonstochastic methods for solving this problem is that there is no known polynomial-time algorithm for it. This problem, as many other optimization problems, is NP-complete. Stochastic methods, however, prove to be an effective alternative.

The algorithm takes as input the number of words needed to produce and the word length. It also takes as input the set of combinatorial constraints that must be satisfied. The first step of the algorithm is to produce a random set of k words. This set is then iteratively modified by decreasing the number of constraint violations at each step. More specifically, two words w_1 and w_2 are chosen from the set that violate at least one of the constraints. (If such words cannot be found, the algorithm terminates.) With a probability θ, called the noise parameter, one of these words is altered by randomly substituting one base. This substitution, of course, does not necessarily improve the set. Alternatively, with probability $1 - \theta$, one of w_1 or w_2 is modified by substitution of one base in a way that maximally decreases the number of conflict violations. The algorithm terminates either when there are no more conflicts in the set of words or when the number of loop iterations has exceeded some maximum threshold.

One drawback of this algorithm is that it may stagnate towards the end in the sense that no improvements to the word set are made after a certain number of steps. This stagnation effect can be overcome by replacing a subset of the words from the set at the point of stagnation with randomly generated words and restarting the algorithm until it reaches the next stagnation. Empirical results prove this technique to be effective. The noise parameter θ is empirically determined to be optimal as 0.2, regardless of the problem instance.

THEORETICAL STUDIES

DNA Sequence Design

The design of DNA sequences optimal for computation can be done with the aid of software, as described in earlier sections. However, some refinement can be added to the process by appealing to theoretical methods. One technique complementary to the algorithmic construction of DNA sequences uses template-based design (Arita & Kobayashi, 2002). The goal, once again, is to create a set S of words of length l over a four-letter alphabet $\Sigma = \{A, C, G, T\}$ that are as dissimilar as possible. The degree of similarity can be measured with the Hamming distance or some other suitable metric. It is also important that melting temperatures for different pairs of strands are in close proximity. For the purposes of this particular template-based design, it is assumed that similarity of melting temperatures can be achieved by standardizing the GC content of strands. That is, all strands are required to contain an equal proportion of G or C nucleotides with respect to the number of A and T nucleotides. This tactic is motivated by a so-called 2–4 rule that says that the melting temperature for a short (14 to 20 base pairs) DNA duplex is roughly equal to two times the number of A-T base pairs, plus four times the number of G-C base pairs.

The template method creates a set of DNA sequences in two separate stages. In the first stage, a template is chosen according to specific criteria. The template is a sequence over the binary alphabet, where the digit 1 indicates the location of either A or T nucleotide and 0 indicates the place of C or G. In the second stage of the design process, an error-correcting code over the binary alphabet is chosen with words of the same length as that of the template. Each code word is used in combination with the template to create a DNA sequence. The "1" locations in the code word mark places where either A or G will be selected in the future. The "0" locations indicate that the choice will be between T and C. Consider an example template 110100 and code word 101010. The choice of the template implies the sequence is $[AT] [AT] [GC] [AT] [GC] [GC]$, where $[AT]$ means that either A or T is present in that location, and similarly for $[GC]$. The choice is made according to the template word: letters in the odd-numbered locations are picked to be A or G, and those in even locations are picked as T or C. The resulting DNA sequence is, hence, $ATGTGC$. This procedure is followed for each code word in order to obtain the resulting set of DNA sequences. Notice that all sequences will have an identical GC content, as determined by the template.

There are two separate tasks involved in this method. One is to find a suitable template, and the other is to find an error-correcting code of length l. The latter of these tasks is a well-studied problem in coding theory, so the remaining challenge is in finding a good template. What is a good template in this case? It is a binary string x of length l that sufficiently differs from its reverse x^R and from any overlap of its concatenation xx, with possibly reversed occurrences: xx^R, x^Rx, or x^Rx^R. More formally, the *mass* of x is defined to be:

$$mass(x) = min\ [\ H(x,\ x^R),\ H_M(x,\ xx),\ H_M(x,\ x^Rx^R),\ H_M(x,\ xx^R),$$
$$H_M(x,\ x^Rx)\ ].$$

In this definition, H is the Hamming distance between two strings of equal lengths. In order to apply a similar measure for strings of different lengths, the measure $H_M(u,v)$ (though not a distance) is defined as the minimum Hamming distance between u and substrings of length $|u|$ in v. What we are looking for is the template x such that $mass(x) = d$. This will ensure that the sequences in the constructed set S differ from other sequences, their concatenations, and reverse complements in at least d locations.

Theoretical analysis methods suggested by Arita and Kobayashi (2002) make it feasible to use an exhaustive search to find an appropriate template, as the search space is significantly reduced. A list of suitable templates is made available by the authors. The number of sequences constructed with this method is limited to the size of the code used. However, it is possible to use multiple templates simultaneously, provided that they are significantly different. A limitation of the template-based sequence design is that it ignores the possibility of strands forming secondary structures.

Different methods of construction are described in Kari et al. (2003b). First, a stronger dissimilarity of sequences in a set K is required — each two subsequences w_1 and w_2 of a fixed length l must satisfy the RC constraint (see the previous section), meaning $H(w_1, wcc(w_2)) \geq d$. At this point, the set K is called $(\theta, H_{d,l})$-bond-free. The number d expresses the degree of dissimilarity — the number of mismatches per each l members of a sequence. Suppose that d mismatches (noncomplementary pairs of nucleotides) are enough to prevent a stable bond between any two (sub)sequences of length l. Then the $(\theta, H_{d,l})$-bond-freedom ensures that the strands in K can form neither (partial) duplexes, nor a stable secondary structure.

One of the most interesting construction methods based on $(\theta, H_{d,l})$-bond-freedom relies on the operation of *subword closure*. Consider a set S of single strands of a fixed length l. Its subword closure $S^\circledast$ is the set of all the possible strands with at least l nucleotides, such that each subsequence of $S^\circledast$ of length l is in S. The following important result holds: if S is $(\theta, H_{d,l})$-bond-free, then so is $S^\circledast$. Moreover, $S^\circledast$ is regular and its description (finite automaton or regular grammar) can be constructed from S in linear time. This description allows a rapid construction of $(\theta, H_{d,l})$-bond-free sets. Assume that we want to construct a $(\theta, H_{d,l})$-bond-free set of strands of a fixed length $k \geq l$. There is an algorithm that at each k steps produces one such strand. Moreover, this construction method also produces *maximal* $(\theta, H_{d,l})$-bond-free sets. Such a set is called maximal if no further sequence can be added without violation of $(\theta, H_{d,l})$-bond freedom. This result can be strengthened as follows: if S is a maximal $(\theta, H_{d,l})$-bond-free set, then so is $S^{\otimes}$. Hence, with respect to the amount of produced strands (of a given length) that are free of all mutual (partial) bonds, this method is provably optimal.

DNA Languages and Their Properties

A set of single-stranded DNA sequences can be viewed as a formal language over a four-letter alphabet $\Delta = \{A, C, G, T\}$. For an introduction to formal language theory, the reader is referred to Hopcroft et al. (2001). Theoretical studies of language properties can shed light on some characteristics of molecular interactions and aid in the design of encoding strategies. Some representative studies include those by Kari et al. (2003) and Jonoska and Mahalingam (2002). The hybridization of two single strands is formalized by defining a function θ, which acts on a sequence over the Δ alphabet to produce the reverse complement of the original string. For example, $\theta(ACCTGACT) = AGTCAGGT$, which corresponds to the fact that the two sequences $ACCTGACT$ and $AGTCAGGT$ would hybridize to form a DNA duplex. While this function attempts to formalize the hybridization process to facilitate theoretical analysis, it has many limitations. It does not take into account base-pair mismatches that can occur during hybridization. The length of the strand is also not taken into account. This function merely reflects the property that if $\theta(u) = v$, then u and v are Watson-Crick complements. The relative simplicity of this function provides several advantages. It allows us to reason about DNA strand interactions theoretically. Consider, for example, the stage of a programmed mutagenic unary counter in which a primer hybridizes to the template. If u is the template and w is the primer, then the template can be written as $u = x\theta(w)y$ for some nonempty strings x and y. A helpful feature of θ is that it happens to be an antimorphism, which is a type of function such that $\theta(uv) = \theta(v)\theta(u)$. This is a useful and well-understood property of functions. Another property of θ is that θ^2 is the identity function. A function for which this is true is called an *involution*. Therefore, θ is an antimorphic involution.

The question addressed in earlier parts of this chapter can be asked again from the theoretical perspective. How can we design DNA sequences that minimize undesirable hybridizations?

First, we need to introduce some notation. An *alphabet* is a finite, nonempty set of symbols. Let Σ be an arbitrary such alphabet. Then Σ^* denotes the set of all words over this alphabet, including the empty word λ. Σ^+ is the same as Σ^*, but without the empty word.

Suppose we want to avoid the type of hybridization shown in Figure 2. A language L is defined (Kari et al., 2003) to be θ-compliant if for all words w in L, and words x and y in Σ^*, we have that $w, x\theta(w)y \in L$ imply $xy = \lambda$. In this definition, Σ is an arbitrary alphabet and θ is taken to be any antimorphic involution. If Σ is taken to be the DNA alphabet Δ, then this property is essentially saying that if two DNA sequences could form a structure such as that shown in the figure, then the unhybridized ends of the longer strand (these are called sticky ends) are of length zero. That is, only the special variant of this structure could occur, as shown by Figure 3.

A language L is said to be θ-nonoverlapping if $L \cap \theta(L) = \varnothing$. A language that is both θ-compliant and θ-nonoverlapping is called strictly θ-compliant. In the context of the DNA alphabet, a strictly θ-compliant language over Δ avoids both of the structures in Figures 2 and 3.

Depending on which type of annealing needs to be avoided, a different language property can be defined. For example, a language is called θ-3'-overhang-free if the following condition holds: $\forall w \in \Sigma^+$, $x, y \in \Sigma^*$ when wx, $\theta(w)y \in L$ implies $xy = \lambda$.

When $\Sigma = \Delta$ and θ is the Watson-Crick antimorphic involution, a language is θ-3'-overhang-free when the structure shown in Figure 4 is avoided. In this DNA structure, a DNA duplex has two overhanging unhybridized ends. These sticky ends are the 3'-ends of the molecules.

Figure 2. A language L *is θ-compliant if it avoids this structure*

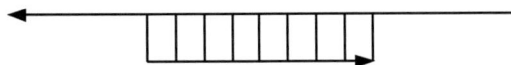

Figure 3. Pattern allowed in a θ-compliant language, but not in a strictly θ-compliant language

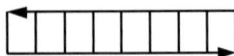

Figure 4. Pattern avoided in a θ-3'-overhang-free DNA language

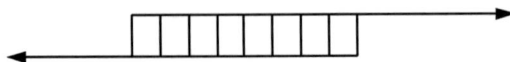

Many other language classes have been similarly defined and their properties studied.

Recently, a more general property of languages, called the bond-free property, was introduced by Kari et al. (2003a). A class $\mathcal{P}$ of languages over Σ is called a *bond-free property of degree 2* if there exist binary word operations $\Diamond_{dn}$ and $\Diamond_{up}$ such that for an arbitrary $L \subseteq \Sigma^*$, $L \in \mathcal{P}$ holds if and only if $\forall w \in \Sigma^+$, $x, y \in \Sigma^*$, $(w \Diamond_{dn} x \cap L \neq \varnothing$, $w \Diamond_{up} y \cap \theta (L) \neq \varnothing)$ implies $xy = \lambda$. The "degree 2" in the name refers to the fact that binding interaction occurs between two DNA strands. To see how this property can generalize other properties, consider the θ-3'-overhang-free property as an example. Let $w \Diamond_{dn} x = \{wx\}$ and $w \Diamond_{up} y = \{yw\}$. Then the definition of bond-free property of degree 2 is instantiated as:

$$\forall w \in \Sigma^+, x, y \in \Sigma^*, \{wx\} \cap L \neq \varnothing, \{yw\} \cap \theta (L) \neq \varnothing \text{ imply } xy = \lambda.$$

This is the same as:

$$\forall w \in \Sigma^+, x, y \in \Sigma^*, wx \in L, yw \in \theta (L) \text{ imply } xy = \lambda.$$

If θ is antimorphic, then $yw \in \theta (L)$ if and only if $\theta(w) \theta (y) \in L$. Note that $xy = \lambda$ only if $x \theta(y) = \lambda$, which means that choosing $w \Diamond_{dn} x = \{wx\}$ and $w \Diamond_{up} y = \{yw\}$ in the definition of bond-free property of degree 2 reduces it to the definition of θ-3'-overhang-free property.

This definition of bond-free property of degree 2 covers many interactions between two DNA strands. Ten various DNA language properties, such as θ-compliance, θ-3'-overhang-freedom and others, were shown to be its special cases. Moreover, it offers some general solutions addressing DNA sequences without undesirable hybridizations.

Assume, for instance, that there is a certain set of DNA sequences. We want to test theoretically (without experiments) whether undesirable bonds of any mentioned type may occur between a pair of sequences in this set. If the set is finite or even *regular*, then the existence of an effective (quadratic time) testing algorithm was proven. In reality, of course, we always deal with finite sets of DNA sequences. The generalization to regular sets, however, might be important. These sets can be very concisely described by means of *regular grammars* (or *finite automata*). Hence, for large sets of DNA strands this concise description allows us to run the algorithms much faster. Also the maximality problem already mentioned in the previous section was addressed. For instance, consider any finite θ-compliant set of DNA sequences. There exists an effective (cubic time) algorithm to decide whether new sequences can be added to this set without loss of θ-compliance.

Suppose we use codes (languages with the property that a catenation of words from a language has a unique factorization over this language) that have the language properties we have described. What may happen during the course of computation is that the properties initially present deteriorate over time. This leads to another area of study, which investigates how bio-operations such as cutting, pasting, splicing, contextual insertion, and deletion affect the various bond-free properties of DNA languages. Invariance under these bio-operations is studied by Kari et al. (2003). This paper also discusses how to add error-detecting capabilities to the codes with these properties. Other studies of coding properties of DNA languages are those by Head (2002) and Hussini et al. (2003). Bounds on the sizes of some other codes with desirable properties that can be constructed are explored by Marathe et al. (2001). Earlier results on code sizes can be found in the work of Baum (1998) and Garzon et al. (1997).

CONCLUSION AND FUTURE TRENDS

This chapter presents an overview of computer science methods that can help reduce errors associated with DNA computing. These methods include software that can find optimal encodings with properties conducive to the reduction of errors, algorithmic tactics to deal with errors, and recent studies of language properties desirable for codes in DNA computing.

It is important to remember that since this computational paradigm is biological in nature, the biochemical methods of dealing with errors are of paramount importance. Indeed, in nature the DNA processes occurring in living organisms only rarely lead to uncorrectable errors, such as carcinogenic mutations. It is remarkable, and almost miraculous, that DNA-related errors in nature are frequently either correctable or inconsequential. Hence, a big question that remains is to understand what error-detecting and error-correcting mechanisms are present in vivo. Since the theory of codes offers a wealth of knowledge in error detection and correction, it is only natural to attempt to apply it to DNA-based languages. More studies of these codes are likely to be carried out in the future.

It is hoped that the strategies described in this chapter can eventually reduce the gap between error rates in nature and those in computational designs carried out in vitro, or at least, explain why the gap is there.

REFERENCES

Adleman, L. (1994). Molecular computation of solutions to combinatorial problems. *Science*, 266, 1021-1024.

Andronescu, M., Dees, D., Slaybaugh, L., Zhao, Y., Cohen, B., Condon, A., & Skiena, S. (2003). Algorithms for testing that sets of DNA words concatenate without secondary structure. *Proceeding of the Eighth International Conference on DNA-Based Computers*, Hokkaido, Japan, June 2002. *Lecture Notes in Computer Science*, 2568, Berlin: Springer, 182-195.

Arita, M., & Kobayashi, S. (2002). DNA sequence design using templates. *New Generation Computing*, 20, 263-277.

Balan, M., Sakthi, Krithivasan, K., & Sivasubramanyam, Y. (2002). Peptide computing: Universality and computing. *Proceedings of the Seventh International Conference on DNA-Based Computers. Lecture Notes in Computer Science,* 2340, Berlin: Springer, 290-299.

Baum, E. B. (1998). DNA sequences useful for computation. *DIMACS Series in Discrete Mathematics and Theoretical Computer Science*, 44, 235-242.

Borer, P. N., Dengler, B., Tinoco, I., Jr., & Uhlenbeck, O. C. (1974). Stability of ribonucleic acid double-stranded helices. *Journal of Molecular Biology,* 86, 843-853.

Braich, R. S., Johnson, C., Rothemund, P. W. K., Hwang, D., Chelyapov, N., & Adleman, L. M. (2001). Solution of a satisfiability problem on a gel-based DNA computer. *Proceedings of the Sixth International Workshop on DNA-Based Computers,* Leiden, The Netherlands, June 2000. *Lecture Notes in Computer Science*, 2054, Berlin: Springer, 27-42.

Faulhammer, D., Cukras, A. R., Lipton, R. J., & Landweber, L. F. (2000). Molecular computation: RNA solutions to chess problems. *Proceedings of the National Academy of Sciences,* 97, 13690-13695.

Feldkamp, U., Banzhaf, W., & Rahue, H., (2001a). A DNA sequence compiler. Poster presented at the *Sixth International Workshop on DNA-Based Computers*, Leiden, The Netherlands, June 2000.

Feldkamp, U., Saghafi, S., Banzhaf, W., & Rahue, H. (2001b). DNA sequence generator: A program for the construction of DNA sequences. *Proceedings of the Seventh International Meeting on DNA-Based Computers*, Tampa, FL, June 2001. *Lecture Notes in Computer Science*, 2340, Berlin: Springer, 23-32.

Freund, R., Păun, Gh., Rozenberg, G., & Salomaa, A. (1999). Watson-Crick finite automata. *Discrete Mathematics and Theoretical Computer Science*, 48, 297-327.

Garzon, M., Deaton, R., Neatheary, P., Murphy, R.C., Franceschetti, D. R., & Stevens, S. E., Jr. (1997 June). On the encoding problem for DNA computing. *Proceedings of the Third Annual DIMACS Workshop on DNA-based Computers*, Philadelphia, PA.

Garzon, M. H., & Oehmen, C. (2001). Biomolecular computation in virtual test tubes. *Proceedings of the Seventh International Meeting on DNA-Based Computers*, Tampa, FL, June 2001. *Lecture Notes in Computer Science*, 2340, Berlin: Springer, 117-128.

Hartemink, J., & Gifford, D. K. (1997). Thermodynamic simulation of deoxyoligonucleotide hybridization for DNA computation. *Proceedings of the Third Annual DIMACS Workshop on DNA-Based Computers*, Philadelphia, PA.

Hartemink, J., Gifford, D. K., & Khodor, J. (1999). Automated constraint-based nucleotide sequence selection for DNA computation. *Biosystems*, 52(1-3), 227-235.

Head, T. (1987). Formal language theory and DNA: An analysis of the generative capacity of recombinant behaviors. *Bulletin of Mathematical Biology*, 49, 737-759.

Head, T. (2002). Relativised code concepts and multi-tube DNA dictionaries, submitted.

Hopcroft, J., Ullman, J., & Motwani, R. (2001). *Introduction to automata theory, languages, and computation* (2nd ed.), Boston: Addison-Wesley.

Hussini, S., Kari, L., & Konstantinidis, S. (2003). Coding properties of DNA languages. *Theoretical Computer Science*, 290, 107-118.

Jonoska, N. & Mahalingam K. (2002). Languages of DNA based code words. *Proceedings of the Ninth International Conference on the DNA-Based Computers*, Madison, WI.

Kari, L., Konstantinidis, S., & Sosík, P. (2003a). On properties of bond-free DNA languages, to appear in *Theoretical Computer Science*, track C.

Kari, L., Konstantinidis, S., & Sosík, P. (2003b). Bond-free DNA languages: Formalizations, maximality and construction methods, to appear in *Int. Journal of Foundations of Comp. Science*.

Kari, L., Konstantinidis, S., Losseva, E., & Wozniak, G. (2003). Sticky-free and overhang-free DNA languages. *Acta Informatica*, 40, 119-157.

Kim, D., Shin, S-Y., Lee, I-H., & Zhang, B-T. (2002). NACST/Seq: A sequence design system with multiobjective optimization. *Proceedings of the Eighth International Conference on DNA-Based Computers*, Hokkaido, Japan, June 2002. *Lecture Notes in Computer Science*, 2568, Berlin: Springer, 242-251.

Liu, Q., Guo, Z., Condon, A., Corn, R. M., Lagally, M. G., & Smith, L. M. (1998). A surface-based approach to DNA computation. *Journal of Computational Biology*, *5*(2), 255-267.

Marathe, A., Condon, A., & Corn, R. M. (2001). On combinatorial DNA word design. *Journal of Computational Biology*, *8*(3), 201-220.

Păun, G. (2000). Computing with membranes. *Journal of Computer and System Sciences*, *61*(1), 108-143.

Roweis, S., Winfree, E., Burgoyne, R., Chelyapov, N. V., Goodman, M.F., Rothemund, P. W. K., & Adleman, L. (1998). A sticker based model for DNA computation. *Journal of Computational Biology*, *5*(4), 615-629.

SantaLucia, J., Jr., Allawi, H. T., & Seneviratne, P. A. (1996). Improved nearest-neighbour parameters for predicting DNA duplex stability. *Biochemistry*, 35, 3555-3562.

Tulpan, D., Hoos, H., & Condon, A. (2003). Stochastic local search algorithms for DNA word design. *Proceedings of the Eighth International Workshop on DNA-Based Computers*, Hokkaido, Japan, June 2002. Lecture notes in *Computer Science*, 2568, Berlin: Springer, 229-241.

Watson, J., Hopkins, N., Roberts, J. W., Steitz, J.A., & Weiner, A. (1987). *Molecular biology of the gene* (4th ed.). Menlo Park, CA: Benjamin Cummings.

Chapter IV

Networks of Evolutionary Processors:
Results and Perspectives

Carlos Martín-Vide, Rovira i Virgili University, Spain

Victor Mitrana, University of Bucharest, Romania, and
Rovira i Virgili University, Spain

ABSTRACT

The goal of this chapter is to survey, in a systematic and uniform way, the main results regarding different computational aspects of hybrid networks of evolutionary processors viewed both as generating and accepting devices, as well as solving problems with these mechanisms. We first show that generating hybrid networks of evolutionary processors are computationally complete. The same computational power is reached by accepting hybrid networks of evolutionary processors. Then, we define a computational complexity class of accepting these networks and prove that this class equals the traditional class NP. In another section, we present a few NP-complete problems and recall how they can be solved in linear time by accepting networks of evolutionary processors with linearly bounded resources (nodes, rules, symbols). Finally, we discuss some possible directions for further research.

INTRODUCTION

A rather well-known architecture for parallel and distributed symbolic processing related to the Connection Machine (Hillis, 1985) and the logic flow paradigm (Errico & Jesshope, 1994) consists of several processors, each of them being placed in a node of a virtual complete graph, which is able to handle data associated with the respective node. Each node processor acts on the local data in accordance with some predefined rules, and then local data becomes a mobile agent that can navigate in the network following a given protocol. Only those data that can pass a filtering process can be communicated to the other processors. This filtering process may require satisfying some conditions imposed by the sending processor, by the receiving processor, or by both of them. All the nodes send their data simultaneously, and the receiving nodes also simultaneously handle all of the arriving messages, according to some strategies (see, e.g., Fahlman et al., 1983; Hillis, 1985).

Starting from the premise that data can be given in the form of words, Csuhaj-Varjú and Salomaa (1997) have introduced a concept called *networks of parallel language processors* with the aim of investigating this concept in terms of formal grammars and languages. Networks of parallel language processors are closely related to grammar systems (Csuhaj-Varjú et al., 1993), and more specifically to parallel communicating grammar systems (Păun & Săntean, 1989). The main idea is that one can place a language-generating device (grammar, Lindenmayer system, etc.) in each node of an underlying graph. Each device rewrites the words existing in the corresponding node, and the words are then communicated to the other nodes. Words can be successfully communicated if they pass some output and input filters. More recently, Csuhaj-Varjú and Salomaa (2003) have introduced networks whose nodes are standard Watson-Crick D0L systems, which communicate to each other either the correct or the corrected words.

In Castellanos et al. (2001), this concept was modified in a way inspired from cell biology. Each processor placed in the nodes of the network is a very simple processor, an evolutionary processor. By an evolutionary processor, we mean a processor that is able to perform very simple operations, namely formal language theoretic operations, that mimic the point mutations in a DNA sequence (insertion, deletion or substitution of a pair of nucleotides). More generally, each node may be viewed as a cell having genetic information encoded in DNA sequences that may evolve by local evolutionary events, namely point mutations. Each node is specialized just for one of these evolutionary operations. Furthermore, the data in each node is organized in the form of multisets of words (each word appears in an arbitrarily large number of copies), and all copies are processed in parallel such that all of the possible events that can take place do so. From the biological point of view, it cannot be expected that the components of any biological organism evolve sequentially or that cell reproduction may be modelled within a sequential approach. Cell state changes are modelled by rewriting rules as in formal grammars. The parallel nature of the cell state changes is modelled by the parallel execution of the symbol rewriting according to the rules applied. Consequently, hybrid networks of evolutionary processors might be viewed as bio-inspired computing models. Obviously, the computational process described here is not exactly an evolutionary process in the Darwinian sense. But the rewriting operations we have considered might be interpreted as mutations, and the filtering process might be viewed as a selection process. Recombination is missing, but it was asserted that evolutionary and functional relationships between genes can be captured by taking only local mutations into consideration (Sankoff et al., 1992).

The mechanisms introduced by Castellanos et al. (2001) are further considered in a series of subsequent works as language-generating devices and their computational power in this respect is investigated. Furthermore, filters based on random-context conditions (Csuhaj-Varjú & Salomaa, 1997) are further generalized in various ways (Castellanos et al., 2001; Castellanos et al., 2003). More precisely, the new filters are based on different types of random-context conditions that seem to be more appropriate for an automated or biological implementation, although this problem is not addressed in the current approach. In the aforementioned papers, the filters of all nodes are defined by the same random-context condition type. Moreover, the rules are applied in the same manner by all of the nodes. These restrictions are discarded in some approaches (Martín-Vide et al., 2003; Margenstern et al., 2004) and for this reason these networks are called hybrid mechanisms.

A similar concept, introduced by Csuhaj-Varjú and Mitrana (2000), is inspired by the evolution of cell populations, which might model some properties of evolving cell communities at the syntactical level. Cells are represented by words that describe their DNA sequences. Informally, at any moment of time, the evolutionary system is described by a collection of words, where each word represents one cell. Cells belong to species and their community evolves according to mutations and divisions that are defined by operations on words. Only surviving (correct) cells are accepted; they are represented by words in a given set, called the *genotype space* of the species.

This feature parallels the natural process of evolution. It is worth mentioning that any recursively enumerable language is a language of a species of an evolutionary system with point mutations of restricted forms. In the aforementioned paper, a connection between Lindenmayer systems (language theoretical models of developmental systems) and evolutionary systems is established; namely, the growth function of any deterministic 0L system can be obtained from the population growth relation of some (deterministic) evolutionary system.

The aforementioned models, besides the mathematical motivation, may also have a biological meaning. Cells always form tissues and organs interacting with each other either directly or via the common environment. In this context, our approach is closely related to "tissue-like" or "neural-like" membrane systems (see Chapter 6 in Păun, 2002). A membrane supercell system (also called a P system) is a theoretical model of the structure and functioning of the cell membrane, which serves as an interface between various inner layers and the cell interior and the exterior environment, within a multicellular structure (Păun, 2002). P systems are distributed parallel computing devices of a biochemical inspiration incorporating complex chemical reactions involving various molecules, catalysts, promoters or inhibitors, electrical charges, etc., and borrowing ideas from Lindenmayer systems, grammar systems, chemical abstract machines, and multiset rewriting models.

The line of research discussed in this chapter lies among a wide range of language-based approaches rooted in molecular biology. Although biologists have long made use of linguistic metaphors in describing processes involving nucleic acids, protein sequences, and cellular phenomena, since the 1980s molecular sequences have been investigated with the methods and tools derived from Chomsky's legacy (Searls, 1988, 1993, 1995; Gheorghe & Mitrana, 2004). Some surveys present various applications of Chomsky grammars or derivatives of them in biology (Searls, 1997, 2002). These approaches, in the context of formal language-based modelling, also led to the emergent area of natural computing, which investigates new computational paradigms rooted in biology (Păun et al., 1998). Treating chromosomes and genomes as languages raises the possibility to generalize and investigate the structural information contained in biological sequences by means of formal language theory methods. It is suggested (Searls, 1995) that both syntactical and functional structure of formal grammars can be modelled by sets of nucleotides and hybridization experiments, respectively.

Chromosomal rearrangements including pericentric and paracentric inversions, intrachromosomal and interchromosomal transpositions, and translocations are modelled as operations on languages (Searls, 1988; Dassow et al., 2001; Yokomori & Kobayashi, 1995). A language-generating mechanism based on the operations suggested by all mutations mentioned earlier is introduced in Dassow et al. (2001), and some properties are studied in Ardelean et al. (2003).

The fundamental mechanism by which genetic material is merged is called *recombination* — DNA sequences are recombined under the effect of enzymatic activities. Head (1987) introduced the *splicing* operation as a language-theoretical approach of the recombinant behaviour of DNA sequences under the influence of restriction enzymes and ligases. According to this approach, a splicing operation consists of cutting two DNA sequences at specified sites, and the first substring of one sequence and the second segment of the other one are then linked at their sticky ends and vice versa. A new type of computability model — called H systems — based on the splicing operations has been considered. Many variants of H systems have been invented and investigated (regulated H systems, distributed H systems, H systems with multisets, etc.). Under certain circumstances, the H systems are computationally complete and universal (Păun et al., 1998). These results suggest the possibility of considering H systems as theoretical models of programmable, universal DNA computers based on the splicing operation.

The bio-operations of gene (un)scrambling in ciliates have been considered as formal operations on strings and languages. First, a computational model based on one intermolecular and one intramolecular operation has been considered (Kari & Landweber, 2000). Another model suggested by the intricate process that transforms the DNA in the micronucleus of some hypotrichous ciliates into that of the macronucleus, based upon three intramolecular operations, has been devised (Ehrenfeucht et al., 2001) and investigated for various properties.

The goal of this chapter is to survey, in a systematic and uniform way, the main results regarding different computational aspects of hybrid networks of evolutionary processors viewed both as generating and accepting devices, as well as solving problems in mechanisms reported so far.

The chapter is organized as follows: in the next section we give the definition of an evolutionary processor and of both generating and accepting hybrid networks of evolutionary processors (GHNEP and AHNEP, respectively). We then discuss the computational power of these mechanisms and some problems regarding the size complexity (number of nodes) of GHNEPs and languages generated by them. We first show that generating hybrid networks of evolutionary processors are computationally complete and give an upper bound for the size complexity of any language L generated by a GHNEP, which is a linear mapping depending on the number of letters appearing in the words of L only. We show that the size complexity is a connected measure and prove that this measure cannot be algorithmically computed for context-free languages. The problem remains open for regular languages; however, we propose an algorithm for deciding whether or not the size of a regular language is equal to 1. In a new section we consider accepting hybrid networks of evolutionary processors and show that they are also computationally complete. Furthermore, we define a computational complexity class of accepting hybrid networks of evolutionary processors and prove that the classical complexity class NP equals this class. In another section we present a few NP-complete problems and recall how they can be solved in linear time by accepting networks of evolutionary processors with linearly bounded resources (nodes, rules, symbols). The chapter ends with a discussion about further research directions and perspectives.

BACKGROUND

We start by summarizing the notions used throughout the chapter. For more details, the reader is referred to Rozenberg and Salomaa (1997). An *alphabet* is a finite and nonempty set of symbols. The cardinality of a finite set A is written as $card(A)$ and the empty set is denoted by $\varnothing$. The set difference of two sets X and Y is denoted by $X \backslash Y$. Any sequence of symbols from an alphabet V is called a *word* over V. The set of all words over V is denoted by V^* and the empty word is denoted by ε. The length of a word x is denoted by $|x|$ while the number of occurrences of a letter a in a word x is denoted by $|x|_a$. Furthermore, for each nonempty word x, we denote by $alph(x)$ the minimal alphabet W such that $x \in W^*$. We denote by w^R the *mirror image* of the word w and by L^R the language of mirror images of all words in L. A morphism from $(V \cup U)^*$ to V^*, which erases all symbols from U and leaves unchanged all symbols from V, is called a *projection* and is denoted by pr_V.

We say that a rule $a \rightarrow b$, with $a,b \in V \cup \{\varepsilon\}$, is a *substitution rule* if both a and b are not ε; it is a *deletion rule* if $a \neq \varepsilon$ and $b = \varepsilon$; and it is an *insertion rule* if $a = \varepsilon$ and $b \neq \varepsilon$. The set of all substitution, deletion, and insertion rules over an alphabet V is denoted by Sub_V, Del_V, and Ins_V, respectively.

Given a rule σ as we have defined, and a word $w \in V^*$, we define the following *actions* of σ on w:

- If $\sigma \equiv a \to b \in Sub_v$, then $\sigma^*(w) = \{ubv : \exists\ u, v \in V^* (w = uav)\}$ or $\sigma^*(w) = \{w\}$
- If $\sigma \equiv a \to b \in Del_v$, then
  - $\sigma^*(w) = \{uv : \exists\ u, v \in V^* (w = uav)\}$ or $\sigma^*(w) = \{w\}$
  - $\sigma^r(w) = \{u : w = ua\}$ or $\sigma^r(w) = \{w\}$
  - $\sigma^l(w) = \{v : w = av\}$ or $\sigma^l(w) = \{w\}$
- If $\sigma \equiv a \to b \in Ins_v$, then $\sigma^*(w) = \{uav : \exists\ u, v \in V^* (w = uv)\}$, $\sigma^r(w) = \{wa\}$, and $\sigma^l(w) = \{aw\}$

$\alpha \in \{*, l, r\}$ expresses the way of applying a deletion or insertion rule to a word at any position ($\alpha = *$), at the left ($\alpha = l$), or at the right ($\alpha = r$) end of the word, respectively. Note that substitution is always applied to any position. For every rule σ, action $\alpha \in \{*, l, r\}$, and $L \subseteq V^*$, we define the α-action of σ on L by $\sigma^\alpha(L) = \cup_{\{w \in L\}} \sigma^\alpha(w)$. Given a finite set of rules M, we define the α-action of M on the word w and the language L by:

$$M^\alpha(w) = \cup_{\{\sigma \in M\}} \sigma^\alpha(w) \text{ and } M^\alpha(L) = \cup_{\{w \in L\}} M^\alpha(w), \text{ respectively.}$$

In what follows, we refer to the rewriting operations defined earlier as *evolutionary operations* since they may be viewed as linguistic formulations of local gene mutations. For two disjoint subsets P and F of an alphabet V and a word over V, we define the predicates:

$$\varphi^{(1)}(w;P,F) \equiv P \subseteq alph(w) \qquad \wedge \qquad F \cap alph(w) = \varnothing$$
$$\varphi^{(2)}(w;P,F) \equiv alph(w) \subseteq P$$
$$\varphi^{(3)}(w;P,F) \equiv P \subseteq alph(w) \qquad \wedge \qquad F \not\subseteq alph(w)$$
$$\varphi^{(4)}(w;P,F) \equiv alph(w) \cap P \neq \varnothing \qquad \wedge \qquad F \cap alph(w) = \varnothing$$

The construction of these predicates is based on *random-context conditions* defined by the two sets P (permitting contexts) and F (forbidding contexts). For every language $L \subseteq V^*$ and $\beta \in \{(1), (2), (3), (4)\}$, we define:

$$\varphi^\beta(L,P,F) = \{w \in L : \varphi^\beta(w;P, F)\}$$

An *evolutionary processor over V* is a tuple (M, PI, FI, PO, FO), where:

- Either $(M \subseteq Sub_v)$ or $(M \subseteq Del_v)$ or $(M \subseteq Ins_v)$. The set M represents the set of evolutionary rules of the processor. As one can see, a processor is "specialized" in one evolutionary operation only.

- $PI, FI \subseteq V$ are the *input* permitting/forbidding contexts of the processor, while $PO, FO \subseteq V$ are the *output* permitting/forbidding contexts of the processor.

An evolutionary processor without filters is said to be *free*. We denote the set of evolutionary processors over V by EP_V.

A *generating hybrid network of evolutionary processors* (a GHNEP, for short) is a 7-tuple $\Gamma = (V, G, N, C_0, \alpha, \beta, x_O)$, where the following conditions hold:

- V is an alphabet.
- $G = (X_G, E_G)$ is an undirected graph with the set of vertices X_G and the set of edges E_G, and each edge is given in the form of a set of two nodes. G is called the *underlying graph* of the network.
- $N: X_G \to EP_V$ is a mapping that associates with each node $x \in X_G$ the evolutionary processor $N(x) = (M_x, PI_x, FI_x, PO_x, FO_x)$.
- $C_0: X_G \to V^*$ is a mapping that identifies the initial configuration of the network. It associates a finite set of words with each node of the graph G.
- $\alpha: X_G \to \{*, l, r\}$; $\alpha(x)$ gives the action mode of the rules of node x on the words associated with that node.
- $\beta: X_G \to \{(1), (2), (3), (4)\}$ defines the type of *input and output filters* of a node. More precisely, for every node, $x \in X_G$, we define the following filters: the input filter is given as $\rho_x(.) = \varphi^{\beta(x)}(.; PI_x, FI_x)$, and the output filter is defined as $\tau_x(.) = \varphi^{\beta(x)}(.; PO_x, FO_x)$. That is, $\rho_x(w)$ ($\tau_x(w)$) indicates whether or not the word w can pass the input (output) filter of x. More generally, $\rho_x(L)$ ($\tau_x(L)$) is the set of words of L that can pass the input (output) filter of x.
- $x_O \in X_G$ is the *output node* of Γ.

An *accepting hybrid network of evolutionary processors* (AHNEP for short) is a 7-tuple $\Gamma = (V, U, G, N, \alpha, \beta, x_I, x_O)$, where:

- V and U are the input and network alphabet, respectively, $V \subseteq U$.
- G, N, α, β, x_O are defined as earlier, and x_I is the *input node* of Γ.

Given a GHNEP Γ, we say that card(X_G) is the size of Γ, denoted by $size(\Gamma)$. If $\alpha(x) = \alpha(y)$ and $\beta(x) = \beta(y)$ for any pair of nodes $x, y \in X_G$, then the network is said to be *homogeneous*. If the set of rules in every node consists of at most one rule, then the network is said to be *elementary*. Furthermore, a network consisting of free evolutionary processors only is said to be *free*. In the theory of networks, some types of underlying graphs are more common (e.g., rings, stars, grids, etc.). Some of the aforementioned papers (Castellanos et al., 2003; Martín-Vide et al., 2003; Csuhaj-Varjú et al., 2003; Margenstern et al., 2004) investigated networks of evolutionary processors that have underlying graphs of these special forms, but special attention is paid to complete graphs. Thus, a GHNEP (AHNEP) is said to be a *star, ring, grid,* or *complete* GHNEP (AHNEP) if its underlying graph is a star, ring, grid, or complete graph, respectively. The star, ring, and complete graph with n nodes is denoted by S_n, R_n, and K_n, respectively. Most of the results presented in the sequel concern complete GHNEPs (AHNEPs). However, these results can be easily extended to GHNEPs (AHNEPs) that have different underlying structures.

A *configuration* of a GHNEP (AHNEP) Γ as defined above is a mapping $C : X_G \rightarrow 2^{V^*}$ which associates a set of words with every node of the graph. A configuration may be understood as the sets of words that are present in any node at a given moment. A configuration can be changed either by an *evolutionary step* or by a *communication step*. When changing by an evolutionary step, each component $C(x)$ of the configuration C is changed in accordance with the set of evolutionary rules M_x associated with the node x and the way of applying these rules $\alpha(x)$. Formally, we say that the configuration C' is obtained in *one evolutionary step* from the configuration C, written as $C \Rightarrow C'$, if and only if:

$$C'(x) = M_x^{\alpha}(x)(C(x)) \text{ for all } x \in X_G.$$

When changing by a communication step, each node processor $x \in X_G$ sends one copy of each word it has that is able to pass the output filter of x to all node processors connected to x and receives all the words sent by any node processor connected with x providing that they can pass its input filter. Formally, we say that the configuration C' is obtained in *one communication step* from configuration C, written as $C \blacktriangleright C'$, only if:

$$C'(x) = (C(x) - \tau_x(C(x))) \cup \cup_{\{x,y\} \in EG} (\tau_y(C(y)) \cap \rho_x(C(y))) \text{ for all } x \in X_G.$$

If Γ is a GHNEP, a *computation* in Γ is a sequence of configurations C_0, C_1, C_2, …, where C_0 is the initial configuration of Γ, $C_{2i} \Rightarrow C_{2i+1}$ and $C_{2i+1} \blacktriangleright C_{2i+2}$ for all $i \geq 0$. By the previous definitions, each configuration C_i is uniquely determined by the configuration C_{i-1}. In other words, any computation is deterministic.

If the sequence is finite, we have a finite computation. The result of any finite or infinite computation is a language that is collected in the output node of the network. For any computation, $C_0, C_1, C_2, \ldots$, all words existing in the output node at some step belong to the language generated by the network. Formally, the *language* generated by Γ is:

$$L_{gen}(\Gamma) = \cup_{s \geq 0} C_s(x_0).$$

If Γ is an AHNEP, the computation of Γ on the input word $w \in V^*$ is a sequence of configurations $C_0{}^w, C_1{}^w, C_2{}^w, \ldots$, where $C_0{}^w$ is the initial configuration of Γ defined by $C_0{}^w(x_I) = w$ and $C_0{}^w(x) = \varnothing$ for all $x \in X_G, x \neq x_I, C_{2i}{}^w \Rightarrow C_{2i+1}{}^w$ and $C_{2i+1}{}^w \blacktriangleright C_{2i+2}{}^w$ for all $i \geq 0$. By the previous definitions, each configuration $C_i{}^w$ is uniquely determined by the configuration $C_{i-1}{}^w$. In other terms, each computation in an AHNEP is deterministic. A computation as such as this immediately halts if one of the following two conditions holds:

(i)   There exists a configuration in which the set of words existing in the output node x_O is nonempty. In this case, the computation is said to be an *accepting computation*.
(ii) There exist two consecutive identical configurations.

In the aforementioned cases the computation is said to be finite. The language accepted by Γ is:

$$L_{acc}(\Gamma) = \{w \in V^* : \text{the computation of } \Gamma \text{ on } w \text{ is an accepting one}\}.$$

COMPUTATIONAL POWER OF GHNEPS

An important and legitimate problem concerns the generative power of GHNEPs. The first result shows that, despite their simplicity, even elementary GHNEPs are computationally complete. This result appears in Csuhaj-Varjú et al. (2003).

Theorem 1. *Any recursively enumerable language can be generated by an elementary, complete GHNEP.*

Proof (Sketch). $G = (N, T, P, S)$ is a phrase-structure grammar in the Geffert normal form (Geffert, 1991), namely with $N = \{S, A, B, C\}$ and rules of the form $S \rightarrow x, x \in (N \cup T)^+$ and $ABC \rightarrow \varepsilon$. We construct an elementary, complete HNEP

$\Gamma = (U, K_s, N, C_0, \alpha, \beta, N_{final})$ that simulates the derivations in G by the so-called *rotate-and-simulate* method. That is, the words found in the nodes are involved into either the rotation of the rightmost symbol (this symbol of the word is moved to the beginning of the word) or a simulation of a rule of P. The constructed network consists of three parts that do not interfere with each other, despite that all nodes of the network are connected with each other.

Each part, consisting of groups of nodes, is dedicated to exactly one type of tasks — that is, the rotation of a symbol, the simulation of a rule $S{\to}x$, or the simulation of the rule $ABC{\to}\varepsilon$. Assume that the rules $S{\to}x{\in}P$ are labelled in a one-to-one manner by *1, 2, ..., m*.

Γ contains a group of four nodes that realize the rotation of a symbol X for each $X{\in}(N\cup T\cup\{\$\})$, where $\$$ is a new symbol representing the left-end marker of the initial word $\$S$. If a word enters the first node of such a group with the aim of rotating the symbol X, then X has to be its rightmost symbol. First, an arbitrary occurrence of X is replaced by Z_X. If the word has no occurrence of X, then it will never leave this node since it cannot pass the output filter. Once the word has the symbol Z_X, it is sent out, but only the second node of the group is able to receive it. In this node symbol X' is appended to the word as its leftmost symbol. Then, the new word leaves the node through the output filter, and it is able to pass only the input filter of the third node. Here, if its rightmost symbol is not Z_X, the word will remain in the node forever. Otherwise, this rightmost symbol is deleted and the word can leave this node. It can pass only the input filter of the last node of the group, where X' is restored into the original symbol X. Thus, symbol X has been rotated.

Γ has a group of nodes that simulate one application of the rule $r: S{\to}x, x = x_1x_2...x_p$ for some $p\geq 1$, and for any rule of the form $r: S{\to}x, 1\leq r\leq m$. The idea behind this construction is the following: the simulation works only for words that enter the first node of the group having S as their rightmost symbol. This occurrence of S is replaced by Y_r and then, by means of p nodes, symbols $x_p^{(r,p)}, x_{p-1}^{(r,p-1)}, ..., x_1^{(r,1)}$ are appended in turn, in this order, to the beginning of the words. After all these symbols have been added, symbol Y_r is deleted (it must be the rightmost symbol). In the obtained words, symbols $x_k^{(r,k)}$ are restored into the original symbols x_k, no matter the order. These words are communicated among other p nodes until all symbols $x_k^{(r,k)}$ are restored. Note that these nodes of the network can interfere neither with the nodes that simulate the application of any other rule nor with the nodes simulating the rotation of any symbol.

We informally present now the simulation of an application of the rule $ABC{\to}\varepsilon$. First, one occurrence of A, B, C is rewritten by W_A, W_B, W_C, respectively. Then, one Y_C is appended to the left end of the word and W_C is deleted, provided that W_C is the rightmost symbol. This process is then resumed for B and A. This means that the initial word was of the form $zABC$ and it was transformed into $Y_A Y_B Y_C z$. The letters Y_A, Y_B, Y_C are then deleted.

Finally, Γ has two more nodes, one node for removing \$, providing \$ is the leftmost symbol, and the output node, which collects the terminal words. The alphabet U of the network and the network size s are given as follows:

$$U = N \cup T \cup \{\$\} \cup \{Z_X, X' : X \in N \cup T \cup \{\$\}\} \cup \{Y_r : 1 \le r \le m\} \cup$$
$$\{W_A, W_B, W_C, Y_A, Y_B, Y_C\} \cup \{x_k^{(r,k)} \mid r: S \to x_1 x_2 \ldots x_{p(r)}, 1 \le k \le p(r)\},$$
$$s = 29 + 2m + 2(p(1) + p(2) + \ldots + p(m)) + 4 card(T)$$

The reader can now formally construct Γ.

As we stated previously, this result holds for elementary, complete, star GHNEPs. To obtain an elementary, complete, star GHNEP generating the same language, we add one more node, $N_{initial}$, defined as follows: the set of rules is empty, $C_0(N_{initial}) = \{S\}$, all the permitting/forbidding condition sets are empty, $\alpha(N_{initial}) = *$ and $\beta(N_{initial}) = (1)$, and connects all the nodes defined earlier to this node in a star graph.

Let L be a language generated by a complete GHNEP. We define $size(L) = \min\{size(\Gamma) : L = L_{gen}(\Gamma)\}$.

The first natural problem concerns the existence of a constant upper bound for the size of any language generated by a GHNEP. The next theorem shows that this is not the case (Castellanos et al., 2004).

Theorem 2. *The measure size is connected; that is, for any $n \ge 1$ there exists a language L_n such that $size(L_n) = n$.*

Proof. We consider the regular language $L_n = a_1^+ a_2^+ \ldots a_n^+$ for any $n \ge 1$. Let Γ be a homogeneous GHNEP with n nodes $x(1), x(2), \ldots, x(n)$ and:

$M_{x(i)} = \{\varepsilon \to a_i, \varepsilon \to a_{i+1}\}, 1 \le i \le n-1$ $M_{x(n)} = \{\varepsilon \to a_n\}$

$PI_{x(i)} = \{a_1, a_2, \ldots, a_i\}, 1 \le i \le n$ $FI_{x(i)} = \{a_{i+1}, a_{i+2}, \ldots, a_n\}, 1 \le i \le n$

$PO_{x(i)} = PI_{x(i)} \cup \{a_{i+1}\}, 1 \le i \le n-1$ $PO_{x(n)} = \emptyset$

$FO_{x(i)} = \emptyset, 1 \le i \le n-1$ $FO_{x(n)} = \{a_1, a_2, \ldots, a_n\}$

$$C_0(x(1)) = \{a_1\} \qquad\qquad C_0(x(i)) = \varnothing,\ 2 \le i \le n$$
$$\alpha(x(i)) = r,\ 1 \le i \le n \qquad\qquad \beta(x(i)) = (1),\ 1 \le i \le n$$

Obviously, $L_{gen}(\Gamma) = L_n$, hence, $size(L_n) \le n$. Let us suppose that $L_n = L_{gen}(\Gamma')$ for some GHNEP Γ' of size at most n-1. By the pigeonhole principle, there exists a node in Γ' in which many symbols are arbitrarily inserted and will eventually be transformed into two distinct symbols, say a_i and a_j, with $1 \le i < j \le n$. It follows that Γ' will also generate words having a_j before a_i, which is contradictory.

The next result, partly presented by Martń-Vide et al. (2003) and Castellanos et al. (2004), is rather surprising since the size of the GHNEP in the proof, hence its underlying structure, does not depend on the number of states of the given automaton. In other words, this structure is common to all regular languages over the same alphabet, no matter the state complexity of the automata recognizing them. Furthermore, all words of the same length are generated simultaneously. On the other hand, the size of a regular language can be bounded by a linear mapping depending on the number of states of the minimum-state automaton recognizing the given language.

Theorem 3. *Given a regular language L accepted by a DFA with n states,* $size(L) \le min\{2card(alph(L))+3,\ card(alph(L))+5,\ 2n+2\}$.

Proof. We give a proof for the second value only. Let $A = (Q, V, \delta, q_0, F)$ be a DFA-accepting L with $card(V) = n$. We construct the following complete GHNEP:

$$\Gamma = (U,\ K_{n+5},\ N,\ C_0,\ \alpha,\ \beta,\ x_O)$$

The alphabet U is defined by $U = V \cup V' \cup Q \cup \{[as] : s \in Q, a \in V\}$, where $V' = \{a' : a \in V\}$. The set of nodes of the complete underlying graph is $\{x_1, x_2, x_3, x_4, x_O\} \cup \{x_a : a \in V\}$, and the other parameters are given in Table 1.

One can easily prove by induction that:

1. $\delta(q_0,x) = s$ with $s \in Q \backslash F$ if and only if $sx \in C_{8|x|}(x_1)$.
2. $\delta(q_0,x) = s$ with $s \in F$ if and only if $sx \in C_{8|x|}(x_1) \cap C_{8|x|}(x_4)$.

Table 1.

Node	M	PI	FI	PO	FO	C_0	α	β
x_1	$\{q{\to}[as] :$ $\delta(q,a)=s\}$	Q	$U\backslash(Q{\cup}V)$	U	$\varnothing$	$\{q_0\}$	$*$	(4)
x_2	$\{\varepsilon{\to}a' :$ $a{\in}V\}$	$\{[as] :$ $a{\in}V, s{\in}Q\}$	$\varnothing$	U	$\varnothing$	$\varnothing$	r	(4)
$x_a,\ a{\in}V$	$\{[as]{\to}s :$ $s{\in}Q\}$	$\{a'\}$	$U\backslash(V{\cup}\{a'\}{\cup}$ $\{[aq] :q{\in}Q\})$	U	$\varnothing$	$\varnothing$	$*$	(4)
x_3	$\{a'{\to}a :$ $a{\in}V\}$	$\{c' :c{\in}V\}$	$U\backslash(V{\cup}V'{\cup}Q)$	U	$\varnothing$	$\varnothing$	$*$	(4)
x_4	$\{q{\to}\varepsilon :$ $q{\in}F\}$	F	$U\backslash(V{\cup}Q)$	U	$\varnothing$	$\{q_0\}$	$*$	(4)
x_0	$\varnothing$	U	$U\backslash V$	U	$\varnothing$	$\varnothing$	$*$	(4)

Therefore, L is exactly the language generated by Γ. Note that the number of symbols is now $3n+(n+1)card(Q)$ while the number of rules is at most $3n+(2n+1)card(Q)$.

A natural problem arises: Given a regular language L, is $size(L)$ algorithmically computable? We are not able to give a complete answer to this problem. However, we can state the findings of Castellanos et al. (2004) in the following theorem:

Theorem 4. *Given a regular language L, one can algorithmically decide whether or not size(L) = 1.*

Proof (Sketch). Clearly, given a regular language $L\subseteq U^*$, $size(L)=1$ if and only if there exist a finite subset E of L and an alphabet $V\subseteq U$ such that exactly one of the next three conditions is satisfied:

$$(1)\ L = EV^*, \qquad (2)\ L = \|\ (E,V^*), \qquad (3)\ L = V^*E, \qquad (*)$$

where $\|$ is the *shuffle* operation defined for two words $x, y{\in} V^*$ by:

$$\| (x,y) = \{x_1 y_1 x_2 y_2 \ldots x_n y_n : n \geq 1, x_i, y_i \in V^*, x = x_1 x_2 \ldots x_n, y = y_1 y_2 \ldots y_n\},$$

and extended to languages by:

$$\| (E,F) = \cup_{x \in E, y \in F} \| (x,y).$$

We start by showing that the first condition can be algorithmically checked. Clearly, a regular language L that can be decomposed as EV^* with E being a finite set, can be accepted by a DFA satisfying the following properties:

1. All of its cycles are actually loops labelled by letters from V in some final states (possibly all).
2. If a final state has a loop, it has one loop for each letter from V and no other edge going out.
3. If a final state has no loop, then for each letter $a \in V$ there is an edge from this state to another final state labelled with a.

We call such a DFA an *ultimately-looped* DFA. The following fact allows us to give a decision procedure for the question "Can a given language L be decomposed in the form EV^* with E being a finite set?"

Fact 1. *A regular language L can be decomposed as EV^* with E being a finite set, if and only if the minimum-state DFA accepting L is ultimately looped.*

The third condition in (*) can also be algorithmically checked since $L = V^*E$ only if $L^R = EV^*$. It remains to show how the second condition in (*) can be algorithmically tested. First, we can algorithmically determine the possible alphabet V by $V = \{a \in U : pr_{\{a\}}(L)$ is infinite$\}$.

The set V can be computed since the projection of a regular language is a regular language and the finiteness problem is decidable for regular languages. Let A be a DFA recognizing L; clearly all edges of A that belong to a cycle are labelled by letters from V. We define the finite set:

$$F = \{w \in L: w \text{ is computed by a cycle-free path in } A\}.$$

Fact 2. *L can be written as $L = \| (E,V^*)$ for some finite set E if and only if $L = \| (F,V^*)$.*

Proof. It suffices to prove the "only if" part of this fact. Let us assume that L $=||\ (E,V^*)$ for some finite set E. As the shuffle operation is associative, it follows that $||\ (L,V^*) = L$. Since $F \subseteq L$ we infer that $||\ (F,V^*) \subseteq L$. For the converse inclusion, take a word $z \in L$; if z can be computed by a cycle-free path in A, then $z \in F$ and we are done. Assume now that $z \in L$ and all $w \in L$ that can be computed with less cycles than z are $||\ (F,V^*)$. By cutting an elementary cycle in the computation of z we get that $z = x_1 y x_2$ such that $y \in V^+$ and $x_1 x_2 \in L$ can be computed with less cycles than z. By our supposition, $x_1 x_2 \in ||\ (F,V^*)$, hence $z \in ||\ (F,V^*)$ as well, which concludes the proof of the fact and of the theorem.

However, a complete answer to the question "Is the size of a regular language computable?" remains an open problem. The same problem is completely solved for context-free languages by showing that the problem considered in Theorem 4 is not decidable for context-free languages (Castellanos et al., 2004).

Theorem 5. *One cannot algorithmically decide whether the size of a context-free language equals 1.*

Proof. The proof is a reduction to Post's correspondence problem (PCP). We take two Post lists over $\{a, b\}$:

$$x = (x_1, x_2, ..., x_p), \text{ and } y = (y_1, y_2, ..., y_p),$$

and construct the languages:

$$L(z) = \{ba^{i(1)}ba^{i(2)} ... ba^{i(k)}c\alpha^{\ i(k)}\alpha^{\ i(k-1)} ... \alpha^{\ i(1)} : k \geq 1, 1 \leq i(j) \leq p, 1 \leq j \leq k\}, z \in \{x, y\}, \text{ and:}$$

$$L_{PCP}(x, y) = (L(x)\{c\}L^R(y) \cap \{w_1 c w_2 c w^R_2 c w^R_1 : w_1, w_2 \in \{a, b\}^+\}).$$

It is known that $L = \{a, b, c\}^* \backslash L_{PCP}(x, y)$ is context-free. We consider the context-free language $L\{\$\}$, where $\$$ is a new symbol and assume that $size(L\{\$\})=1$. Since $\$ \in L\{\$\}$ it follows that $L\{\$\} = \{a, b, c\}^*\{\$\}$; hence, the given instance of the PCP has no solution. Conversely, if $PCP(x, y)$ has no solution, then $L\{\$\}$ is $\{a, b, c\}^*\{\$\}$, and $size(L\{\$\}) = 1$. In conclusion, since the PCP is not decidable, one cannot decide whether or not $size(L\{\$\}) = 1$.

As an immediate consequence, we state:

Corollary 1. *The measure size is not computable on the family of context-free languages.*

It is well-known that each linear grammar can be transformed into an equivalent linear grammar with rules of the form $A \to aB, A \to Ba, A \to \varepsilon$ only; hence, the proof of Theorem 3 remains valid for linear languages as well. Moreover, the statement remains valid for GHNEPs with other types of underlying structure.

Theorem 6. *Any regular and linear language L over an alphabet with n symbols can be generated by a complete/star/ring GHNEP whose size depends linearly on n, only.*

A natural problem arises: Is it possible to give a similar characterization of other families of languages in the Chomsky hierarchy? Surprisingly enough, the answer is affirmative even for the class of recursively enumerable languages. The basic idea of the proof of the next theorem which appears in (Csuhaj-Varjú et al., submitted) is the same as that of Theorem 1.

Theorem 7. *Any recursively enumerable language over an alphabet V can be generated by a complete HNEP of size 28+3card(V).*

This last result suggests the possibility of constructing a "universal" GHNEP with a fixed underlying structure for all recursively enumerable languages over a given alphabet. The minimal size of a complete or star GHNEP generating an arbitrary recursively enumerable language over a fixed alphabet remains to be further investigated. However, we can state:

Theorem 8.

1. *The language generated by any GHNEP of size 1 is regular.*
2. *There exist non-context-free languages that can be generated by complete, homogeneous GHNEPs of size 2.*
3. *There exist nonrecursive languages that can be generated by complete or star GHNEPs of size 28.*

Table 2.

Node	M	PI	FI	PO	FO	C_0	α	β
N_1	$\{\varepsilon \to a,$ $\varepsilon \to b,$ $\varepsilon \to X\}$	∅	∅	$\{c\}$	∅	$\{\varepsilon\}$	*	*(l)*
N_2	$\{X \to c\}$	$\{X\}$	∅	∅	∅	∅	*	*(l)*

4. *The family of languages generated by complete or star GHNEPs having no deletion node coincides with the family of context-sensitive languages.*

Proof.

1. If the network has only one node, say x, that is a deletion or a substitution node, then the generated language is finite. Let us consider that the node is an insertion node; we assume that it contains the rules $\varepsilon \to a_i, 1 \le i \le n$, for some $n \ge 1$. If $\alpha(x)$ is l or r, then the generated language is $\{a_1, a_2, ..., a_n\}^* C_0(x)$ or $C_0(x)\{a_1, a_2, ..., a_n\}^*$, respectively. If $\alpha(x) = *$, then the generated language is $\shuffle(\{a_1, a_2, ..., a_n\}^*, C_0(x))$.

2. Ralf Stiebe communicated to us (2003) a complete homogeneous GHNEP of size 2 generating a non-context-free language, which is presented below in a simplified version. The alphabet of the network is $\{a, b, c, X\}$ and the two nodes are N_1, N_2. Either of them can be considered the output node. We consider N_1 as the output node. The other parameters are given in Table 2. The nonempty words in the processor N_1 at some moment may contain one, two, or all three letters in the alphabet $\{a, b, c\}$. While the word is in N_1, many as or bs can arbitrarily be inserted. As soon as X is inserted, the word leaves N_1 and enters N_2 where X is replaced by c. Now, the words returns to N_1 and the process resumes. Let L be the language generated by this GHNEP of size 2. We now consider the language $L' = L \cap a^+b^+c^+ = \{a^nb^mc^p : n, m \ge p \ge 1\}$. We argue that this language is not context-free. To this aim we use Ogden's lemma (Ogden, 1968). We assume the contrary and take the word $a^nb^nc^n$ and distinguish all the positions occupied by c (all the occurrences of c). Since for any decom-

position of a sufficiently long word $z = uvwxy \in L'$, the following two conditions must be satisfied:

   (i)   v or x contains at least one distinguished position (one occurrence of c)

   (ii) $uv^iwx^iy \in L'$ for any $i \geq 1$

Here we get a contradiction. Therefore, either L is not context-free.

3.   It is known that there exist nonrecursive languages over the one-letter alphabet. We consider a phrase-structure grammar in the Geffert normal form generating such a language L and construct the complete GHNEP from the proof of Theorem 7, but removing the symbol $ together with the node having the role of deletion of $, and replacing the output node with another existing node. Let us denote this GHNEP of size 28 by Γ. Then, the intersection of the language generated by Γ with the language {a}* is exactly L. Therefore, $L_{gen}(\Gamma)$ is nonrecursive.

4.   By a standard proof one can get that if Γ is a GHNEP having no deletion node, then $L_{gen}(\Gamma) \in$ NSPACE(n).

The smallest size of a non-context-free language is 2. Which is the smallest size of a non-context-sensitive language? What about a nonrecursive language?

We finish this section by pointing out an interesting result from Stiebe (2004), which states that any recursively enumerable language can be generated by a GHNEP with an undirected underlying graph with 58 nodes. In other words, if the complete (ring, star) structure of a GHNEP generating a recursively enumerable language L is discarded, then the size of L can be bounded by a constant. Clearly, by Theorem 2, such a result cannot hold for complete GHNEPs. What about star or ring GHNEPs?

COMPUTATIONAL
COMPLEXITY OF AHNEPS

A nondeterministic *Turing machine* is a construct:

$$T = (Q, V, U, \delta, q_0, B, F),$$

where Q is a finite set of states, V is the input alphabet, U is the tape alphabet, $V \subset U$, q_0 is the initial state, $B \in U \backslash V$ is the "blank" symbol, $F \subseteq Q$ is the set of final states, and δ is the transition mapping, $\delta : (Q \backslash F) \times U \rightarrow 2^{Q \times U \times \{R,L\}}$. The variant of a Turing machine we use in this chapter can be described intuitively as follows: it has a tape divided into cells that may store symbols from U (each cell may store exactly one symbol from U). The tape is semi-infinite; it is bounded to the left (there is a leftmost cell) and unbounded (arbitrarily long) to the right. The machine has a central unit that can be in a state from a finite set of states, and a reading/writing tape head that can scan, in turn, the tape cells. This head cannot go to the left-hand end of the tape. The input word is a word over V and is stored on the tape starting with the leftmost cell, and all other tape cells contain the symbol B.

Initially, the tape head is on the leftmost cell and the central unit is in the state q_0. The machine performs moves, which depend on the contents of the cell currently scanned by the tape head and the current state of the central unit. A move consists of the following actions: one changes the state, writes a symbol from U on the current cell, and moves the tape head one cell either to the left (provided that the cell scanned was not the leftmost one) or to the right. An input word is *accepted* if after a finite number of moves the Turing machine enters a final state.

An *instantaneous description* (ID for short) of a Turing machine T as described earlier is a word over $(U\setminus\{B\})^*Q(U\setminus\{B\})^*$. Given an ID uqv, this means that the tape contents are uv followed by an infinite number of cells containing the blank symbol B (note that all cells to the right of the first cell containing the blank symbol contain blank symbols only), the current state is q, and the symbol currently scanned by the tape head is the first symbol of v (u is the tape contents in the left-hand side of the current cell), provided that $v \neq \varepsilon$, or B, otherwise. We define the relation $\models$ on the set of all IDs of a Turing machine T in the following way:

$uqXv \models uYpv$ iff $(p,\ Y,\ R) \in \delta(q,\ X),\ u,\ v \in (U\setminus\{B\})^*$
$uq \models uYp$ iff $(p,\ Y,\ R) \in \delta(q,\ B),\ u \in (U\setminus\{B\})^*$
$uZqXv \models upZYv$ iff $(p,\ Y,\ L) \in \delta(q,\ X),\ u,\ v \in (U\setminus\{B\})^*$
$uZq \models upZY$ iff $(p,\ Y,\ L) \in \delta(q,\ B),\ u \in (U\setminus\{B\})^*$

Note that the tape head can never write B. The language accepted by the Turing machine T is defined by:

$$L(T) = \{w \in V^* : q_0 w \models^* uqv,\ q \in F\},$$

where $\models^*$ is the reflexive and transitive closure of the relation $\models$.

The reader is referred to Hartmanis et al. (1965) and Hartmanis and Stearns (1965) for the classical time and space complexity classes defined on the standard computing model of the Turing machine. We define some computational complexity measures by using AHNEP as the computing model following Margenstern et al. (2004). To this aim, we consider an AHNEP Γ and the language L accepted by Γ. The *time complexity* of the accepting computation $C_0^{(x)}, C_1^{(x)}, C_2^{(x)}, ..., C_m^{(x)}$ of Γ on $x \in L$ is denoted by $Time_\Gamma(x)$ and equals m.

The time complexity of Γ is the partial function from $\mathbf{N}$ to $\mathbf{N}$,

$$Time_\Gamma(n) = \max\{Time_\Gamma(x) : x \in L_{acc}(\Gamma), |x| = n\}.$$

For a function $f: \mathbf{N} \rightarrow \mathbf{N}$ we define:

$$\mathbf{Time}_{\text{AHNEP}}(f(n)) = \{L : L = L_{acc}(\Gamma) \text{ for an AHNEP } \Gamma \text{ with } Time_\Gamma(n) \leq f(n)$$
for some $n \geq n_0\}$.

Moreover, we write:

$$\mathbf{PTime}_{\text{AHNEP}} = \cup_{k \geq 0} \mathbf{Time}_{\text{AHNEP}}(n^k).$$
$$\mathbf{ExpTime}_{\text{AHNEP}} = \cup_{\{x=2^k, k \geq 0\}} \mathbf{Time}_{\text{AHNEP}}(2^x).$$

Now we recall a result from Margenstern et al. (2004) that makes possible a connection between the complexity classes defined on Turing machines and those defined on AHNEPs. The proof is omitted due to space reasons.

Proposition 1. *For any Turing machine M recognizing a language L, there exists an AHNEP Γ accepting the same language L. Furthermore, $Time_\Gamma(n) \leq 22T_M(n)$.*

It is worth mentioning that the underlying graph of Γ is the complete graph K_p, where:

$$p = 18 + 7(card(U)-1) + card(Q) + (card(U)-1)^2 + 2card(Q)(card(U)-1).$$

As one can see, the number of nodes of Γ is bounded by a quadratic function depending on the number of states and symbols of M. It is useful to note that also the total number of symbols used by Γ in the simulation is bounded by a cubic function depending on the number of states and symbols of M. More precisely, the number of symbols used by Γ is:

$$4\,card(Q)(card(U)-1)^2+2(card(U)-1)^2+2\,card(Q)(card(U)-1)+10(card(U)-1)+card(Q).$$

Theorem 9.

1. $\mathbf{NP = PTime}_{AHNEP}.$
2. $\mathbf{NEXPTIME = ExpTime}_{AHNEP}.$

Proof. We give a proof for the first equality. Let L be a language accepted by a nondeterministic Turing machine M with k tapes such that for each $x \in L$, $|x| = n$, M can accept x in no more than $p(n)$ moves. We write this as $T_M(n) \leq p(n)$. By the well-known results regarding tape compression, we can construct a Turing machine M' with one tape only, such that $T_{M'}(n) \leq p^2(n)/22$. Now, by the previous proof, we construct an AHNEP Γ such that $L(M') = L_{acc}(\Gamma)$ and $Time_\Gamma(n) \leq 22T_{M'} \leq p^2(n)$.

Conversely, let L be a language accepted by an AHNEP Γ in polynomial time $p(n)$. We construct a nondeterministic Turing machine M as follows:

1. M has a finite set of states associated with each node of Γ. This set is divided into disjoint subsets such that each filter (input or output) and each rule has an associated subset of states.
2. M chooses nondeterministically a copy of the input word from those existing in the initial node of Γ (this word is actually on the tape of M in its initial ID) and follows its itinerary through the underlying network of Γ. Let us suppose that the contents of the tape of M is z; M works according to the following strategy labelled by (*):

   (i) When M enters a state from the subset of states associated to a rule $a \rightarrow b$, it applies this rule to an occurrence of a in z, if any, nondeterministically chosen. If z does not contain any occurrence of a, then M blocks the computation.
   (ii) When M enters a state from the subset of states associated to a filter, it checks whether z can pass that filter. If z does not pass it, M blocks the computation. Clearly, M checks first the condition of the current

the computation. Clearly, M checks first the condition of the current node (sending node) output filter and then the condition of the receiving node input filter (which becomes the current node).

(iii)   As soon as M has checked the input filter condition of the output node of Γ, it accepts its input word.

It is rather obvious that M accepts L. If the input word w in the initial node of Γ is in L, then there exists a computation in Γ of time complexity $O(p(|w|))$. Since in any evolutionary step one inserts at most one letter, the length of z in (*) is at most $p(|w|)+|w|$. Clearly, each step (i) and (ii) of (*) can be accomplished in time $O(|z|)$. Therefore, w is accepted by M in $O(p^2(|w|))$ and the proof is complete.

SOLVING HARD PROBLEMS WITH AHNEPS

AHNEPs may be used for solving problems in the following way. For any instance of the problem, the computation in the associated AHNEP is finite. In particular, this means that there is no node processor specialized in insertions. In the initial configuration, the input node contains an arbitrarily large number of copies of a word whose length depends linearly on the input size. Actually, all solutions presented in the sequel can be modified such that the input node contains an arbitrarily large number of copies of the empty word. If the problem is a decision problem, then at the end of the computation, the output node provides all solutions of the problem encoded by words, if any; otherwise, this node will never contain any words. If the problem requires a finite set of words, this set will be in the output node at the end of the computation. In other cases, the result is collected by specific methods, which are indicated for each problem.

Despite their simplicity, these mechanisms are able to solve hard problems in polynomial time. We shall see in the sequel some well-known NP-complete problems can be solved in linear time.

In Castellanos et al. (2003), a solution to an NP-complete problem, namely the *"three-colorability problem"*, is proposed. This problem is to decide whether each vertex in an undirected graph can be colored by using three colors (say, red, blue, and green) in such a way that after coloring, no two vertices connected by an edge have the same color.

Theorem 10. *The "three-colorability problem" can be solved in $O(m+n)$ time by a complete homogeneous AHNEP of size $7m+2$, where n is the number of vertices and m is the number of edges of the input graph.*

Proof. Let G = ({$1, 2, ..., n$}, {$e_1, e_2, ..., e_m$}) be a graph and assume that e_t = {$i(t), j(t)$}, $1 \le i(t) < j(t) \le n$, $1 \le t \le m$. We consider the alphabet:

$$U = V \cup V' \cup T \cup \{X_1, X_2, ..., X_{m+1}\},$$

where $V = \{b_1, r_1, g_1, ... b_n, r_n, g_n\}$, $T = \{a_1, a_2, ..., a_n\}$. Here V' is the primed copy of V—that is, the set formed by the primed copies of all letters in V. We construct the complete homogeneous AHNEP of size $7m+2$ having the nodes (the action mode of the rules in any node is * and the type of any filter is *(1)*):

$$N(x_1) = (\{a_i \rightarrow b_i, a_i \rightarrow r_i, a_i \rightarrow g_i : 1 \le i \le n\}, T \cup \{X_1\}, \varnothing, \varnothing, T).$$

The word in the input node is $a_1 a_2 ... a_n X_1$ while the further nodes are defined by:

$$N(x(e_t)^{(Z)}) = (\{Z_{i(t)} \rightarrow Z'_{i(t)}\}, \{X_t\}, U \backslash (V \cup \{X_t\}, \{Z'_{i(t)}\}, \varnothing), Z \in \{b, r, g\},$$
$$N(<x(e_t)^{(b)}>) = (\{r_{j(t)} \rightarrow r'_{j(t)}, g_{j(t)} \rightarrow g'_{j(t)}\}, \{X_t, b'_{i(t)}\}, \varnothing, \{r'_{j(t)}, g'_{j(t)}\}, \varnothing),$$
$$N(<x(e_t)^{(r)}>) = (\{b_{j(t)} \rightarrow b'_{j(t)}, g_{j(t)} \rightarrow g'_{j(t)}\}, \{X_t, r'_{i(t)}\}, \varnothing, \{b'_{j(t)}, g'_{j(t)}\}, \varnothing),$$
$$N(<x(e_t)^{(g)}>) = (\{r_{j(t)} \rightarrow r'_{j(t)}, b_{j(t)} \rightarrow b'_{j(t)}\}, \{X_t, g'_{i(t)}\}, \varnothing, \{r'_{j(t)}, b'_{j(t)}\}, \varnothing),$$
$$N(x(e_t)) = (\{r'_{i(t)} \rightarrow r_{i(t)}, g'_{i(t)} \rightarrow g_{i(t)}, g'_{i(t)} \rightarrow g_{i(t)}, r'_{j(t)} \rightarrow r_{j(t)},$$
$$b'_{j(t)} \rightarrow b_{j(t)}, g'_{j(t)} \rightarrow g_{j(t)}, X_t \rightarrow X_{t+1}\}, \{X_t\}, \{r_{j(t)}, g_{j(t)}, b_{j(t)}, r_{i(t)},$$
$$g_{i(t)}, b_{i(t)}\} \cup T, \{X_{t+1}\}, U \backslash V),$$
for all $1 \le t \le m$, and $N(x_O) = (\{X_{m+1} \rightarrow \varepsilon\}, \{X_{m+1}\}, U \backslash V, \varnothing, V)$.

For the first $2n$ steps, out of which n steps are communication steps when nothing is actually communicated, the words will remain in x_1 until no letter in T appears in them anymore. When this process is finished, the obtained words encode all possible ways of coloring the vertices, satisfying the requirements of the problem or not. Now, for each edge e_t, our AHNEP keeps only those words that encode a colorability satisfying the condition for the two vertices of e_t. This process is completed by means of the nodes $x(e_t)^{(b)}$, $x(e_t)^{(r)}$, $x(e_t)^{(g)}$, $<x(e_t)^{(b)}>$, $<x(e_t)^{(r)}>$, $<x(e_t)^{(g)}>$, and finally $x(e_t)$ in 8 steps. It is clear that computation on the input word $a_1 a_2 ... a_n X_1$ is finite. Moreover, the given instance has a solution if and only if the language of the node x_O is nonempty. As one can see, the overall time of a computation is $8m+2n$.

We finish the proof by mentioning that the total number of rules is $16m+3n+1$. In conclusion, all parameters of the network are of $O(m+n)$ size.

It is rather interesting that the underlying graph of the AHNEP does not depend on the number of nodes of the given instance of the problem. In other words, the same underlying structure may be used for solving any instance of the three-colorability problem having the same number of edges but any number of nodes.

This construction can be modified in the aim of starting from an infinite number of copies of the empty word only. To this end, we need $n+1$ additional nodes (the node x_1 renamed by x_0). However, the action mode of the rules in these nodes is not $*$ anymore but r.

$$N(x_1) = (\{\varepsilon \rightarrow a_1\}, \varnothing, \varnothing, \varnothing, \varnothing),$$
$$N(x_i) = (\{\varepsilon \rightarrow a_i\}, T_i, U \backslash T_i, \varnothing, \varnothing), \text{ for all } 2 \leq i \leq n,$$
$$N(x_{n+1}) = (\{\varepsilon\ X_1\}, T, U \backslash T, \varnothing, \varnothing).$$

For each $2 \leq i \leq n$, we denoted $T_i = \{a_1, a_2, ..., a_{i-1}\}$. Now, the computation starts with the empty word in x_1 where the word a_1 is formed and sent out. Now, step-by-step, the word $a_1 a_2 ... a_n$ is constructed in x_n. This phase in which $a_1 a_2 ... a_n$ is the unique word produced in $2n$ steps. Now, this word enters x_{n+1} where it receives X_1 in its right end. From now on, the computation follows the description given previously.

Castellanos et al. (2001) presents a linear solution for an NP-complete problem, namely the *bounded Post correspondence problem*. The bounded Post correspondence problem (BPCP) (Constable et al., 1974; Garey & Johnson, 1979) is a variant of Post's correspondence problem, which is known to be unsolvable in the unbounded case (Garey & Johnson, 1979). An instance of the BPCP consists of an alphabet V, two lists of words over V:

$$x = (x_1, x_2, ..., x_n) \qquad \text{and} \qquad y = (y_1, y_2, ..., y_n)$$

and $K \leq n$. The problem asks whether or not a sequence $i(1), i(2), ..., i(k), k \leq K$, of positive integers exists, each between 1 and n, such that:

$$x_{i(1)} x_{i(2)} \cdots x_{i(k)} = y_{i(1)} y_{i(2)} \cdots y_{i(k)}.$$

Theorem 11. *Let* $x = (x_1, x_2, ..., x_n)$ *and* $y = (y_1, y_2, ..., y_n)$ *be two Post lists and* $K \leq n$. *This instance of the bounded PCP can be solved by a complete AHNEP in size and time linearly bounded by the product of K and the length of the longest word of the two Post lists.*

Proof. Let $x = (x_1, x_2, ..., x_n)$ and $y = (y_1, y_2, ..., y_n)$ be two Post lists over the alphabet $V = \{a_1, a_2, ..., a_m\}$ and $K \leq n$. Let:

$$s = K \cdot \max(\{|x_j| : 1 \leq j \leq n\} \cup \{|y_j| : 1 \leq j \leq n\}).$$

Consider a new alphabet:

$$U = \cup_{i=1}^{m} \{a_i^{(1)}, a_i^{(2)}, ..., a_i^{(s)}\} = \{u_1, u_2, ..., u_{sm}\}.$$

For each $w = i(1)i(2)...i(j) \in \{1, 2, ..., n\}^{\leq K}$ (the set of all sequences of length, at most K, formed by integers between 1 and n), we define the word:

$$x^{(w)} = x_{i(1)}x_{i(2)} \cdots x_{i(j)} = a_{t(1)}a_{t(2)} \cdots a_{t(p(x))}.$$

We now define a one-to-one mapping $\xi: V^* \to U^*$ such that for each sequence w, $\xi(x^{(w)})$ does not contain two occurrences of the same symbol from U. We may take:

$$\xi(x^{(w)}) = a_{t(1)}^{(1)}a_{t(2)}^{(2)} \cdots a_{t(p(w))}^{p(w)}$$

The same construction applies to the words in the second Post list y. We define:

$$F = \{\xi(x^{(w)}) \xi(y^{(w)}) : w \in \{1, 2, ..., n\}^{\leq K}\} = \{z_1, z_2, ..., z_l\}$$

and assume that $z_j = u_{j,1}u_{j,2}... u_{j,r(j)}$, $1 \leq j \leq l$, where $|z_j| = r(j)$. By the construction of F, no letter from U appears within any word in F for more than two times. Furthermore, if each letter of $z = \xi(x^{(w)})\xi(y^{(w)}$ appears twice within z, then w represents a solution of the given instance.

We are now ready to define the AHNEP, which computes all of the solutions of the given instance. It is an AHNEP of size $2sm+2$ with the alphabet:

$$W = U \cup \hat{U} \cup \tilde{U} \cup \{X\} \cup \{X_d^{(c)} : 1 \leq c \leq l, 2 \leq d \leq |z_c|\},$$

where $\tilde{U} = \{\tilde{u} : u \in U\}$ (the other set, namely $\hat{U}$, which forms the AHNEP alphabet is defined similarly), and the nodes $1, 2, ..., 2sm+2$ are defined by:

- $M_f = \{\varepsilon \to u_f\}$,
- $FI_f = \{X_d^{(c)} : 2 \leq d \leq |z_c|, 1 \leq c \leq l$ such that $u_f \neq u_{c,d}\} \cup \hat{U} \cup \tilde{U}$,

- $PI_f = FO_f = PO_f = \emptyset$,
- $\alpha(f) = r$, $\beta(f) = (1)$,

for all $1 \leq f \leq sm$,

- $M_{sm+1} = \{X \rightarrow X_2^{(c)}, X_{|z_c|} \rightarrow u_{c,1} : 1 \leq c \leq l\} \cup \{X_d^{(c)} \rightarrow X_{d+1}^{(c)} : 1 \leq c \leq l, 2 \leq d \leq |z_c|-1\}$,
- $FI_{sm+1} = \hat{U} \cup \tilde{U}$,
- $PI_{sm+1} = FO_{sm+1} = PO_{sm+1} = \emptyset$,
- $\alpha(sm+1) = *$, $\beta(sm+1) = (1)$,

and

- $M_{sm+d+1} = \{u_d \rightarrow \tilde{u}_d, u_d \rightarrow \hat{u}_d\}$,
- $FI_{sm+d+1} = \{X_g^{(c)} : 2 \leq g \leq |z_c|, 1 \leq c \leq l\} \cup \{\tilde{u}_d, \hat{u}_d\}$,
- $PI_{sm+d+1} = FO_{sm+d+1} = \emptyset$,
- $PO_{sm+d+1} = \{\tilde{u}_d, \hat{u}_d\}$,
- $\alpha(sm+d+1) = *$, $\beta(sm+d+1) = (1)$,

for all $1 \leq d \leq sm$.

Finally,

- $M_{2sm+2} = \emptyset$,
- $PI_{2sm+2} = PO_{2sm+2} = \emptyset$,
- $FI_{2sm+2} = FO_{2sm+2} = W\backslash(\hat{U} \cup \tilde{U})$,
- $\alpha(2sm+2) = *$, $\beta(2sm+2) = (1)$.

The input node is $sm+1$, the input word is X, and the output node is $2sm+2$.

There are some considerations about the computing mode of this AHNEP. It is easy to note that in the first stage of a computation only the processors 1, 2, ..., $sm+1$ are active. Since the input filter of the others contains all symbols of the form $X_g^{(c)}$, they remain inactive until one word from F is produced in the node $sm+1$. First let us explain how an arbitrary word $z_j = u_{j,1}u_{j,2}... u_{j,r(j)}$ from F can be obtained in the node $sm+1$. One starts by applying the rules $X \rightarrow X_2^{(j)}$, $1 \leq j \leq l$, in the node $sm+1$. The words $X_2^{(j)}$, $1 \leq j \leq l$, obtained in the node $sm+1$ as an effect of an evolutionary step are sent to all the other processors, but for each of these words there is only one processor that can receive it. For instance, the word $X_2^{(c)}$ is accepted only by the node processor f, $1 \leq f \leq sm$, with $u_f = u_{c,2}$. In the next evolutionary step, the symbol $u_{j,2}$ is added to the right-hand end of the word $X_2^{(j)}$ for all $1 \leq j \leq l$. Now, we must complete a communication step. All the words $X_2^{(j)}u_{j,2}$ can pass the output filters of the node processors where they were obtained, but the node processor $sm+1$ is the only one that can receive them. Here the lower subscripts of the symbol X

are increased by one and the process used earlier is resumed in the aim of adjoining a new letter. This process does not apply to a word $X_r^{(j)} u_{j,2} \ldots u_{j,r}$ anymore, if and only if $r = |z_j|$, where $X_r^{(j)}$ is replaced by $u_{j,1}$ resulting in the word z_j. By these considerations, we infer that all the words from F are produced in the node $sm+1$ in $2s$ steps.

Another stage of the computation checks the number of occurrences of any letter within any word obtained in the node $sm+1$, as soon as the word contains only letters in U. This stage is accomplished as follows. Such a word enters a node $sm+d+1$ for some $1 \le d \le sm$, where two situations may appear. If the word has two occurrences of the symbol u_d, these occurrences are replaced by $\tilde{u}_d$ and $\hat{u}_d$, respectively. In this way, after two evolutionary steps and a communication one, the string is allowed to leave this node. If the two occurrences are replaced by the same letter, it remains in $sm+d+1$ forever. Since the permitting condition of the output filter of $sm+d+1$ requires the presence of both $\tilde{u}_d$ and $\hat{u}_d$, any string with one occurrence of u_d only entering this node cannot leave it anymore. Any string that will later be obtained from this string cannot enter again the node $sm+d+1$. After at most $4s$ steps, the computation stops and the node $2sm+2$ has all the shortest words that were produced from those words in F having two occurrences of any letter. As we have seen, these words encode solutions of the given instance of BPCP. If the given instance of BPCP has no solution, the computation stops after $6s$ steps by the second halting condition.

In the sequel, following the descriptive format for three NP-complete problems presented by Head et al. (1999), we present a solution to the *common algorithmic problem*. The three problems are:

1. The *maximum independent set* problem: given an undirected graph $G = (X, E)$, where X is the finite set of vertices and E is the set of edges given as a family of sets of two vertices, find the cardinality of a maximal subset (with respect to inclusion) of X, which does not contain both vertices connected by any edge in E.
2. The *vertex cover* problem: given an undirected graph, find the cardinality of a minimal set of vertices such that each edge has at least one of its extremes in this set.
3. The *satisfiability* problem: for a given set P of Boolean variables and a finite set U of clauses over P, does a truth assignment for the variables of P exist satisfying all the clauses of U?

For detailed formulations and discussions about their solutions, the reader is referred to Garey and Johnson (1979).

These problems can be viewed as special cases of the following algorithmic problem, called the common algorithmic problem (CAP) in Head et al. (1999): let S be a finite set and F be a nonempty family of subsets of S. Find the cardinality of a maximal subset of S that does not include any set belonging to F. The sets in F are called *forbidden sets*. We say that (F, S) is an *(card(S), card(F))* instance of CAP.

Let us show how the three problems we have discussed can be obtained as special cases of CAP. For the first problem, we just take $S = X$ and $F = E$. The second problem is obtained by letting $S = X$ and F contain all sets $o(x) = \{x\} \cup \{y \in X : \{x, y\} \in E\}$. The cardinality one looks for is the difference between the cardinality of S and the solution of the CAP. The third problem is obtained by letting $S = P \cup P'$, where $P' = \{p' : p \in P\}$, and $F = \{F(C) : C \in U\}$, where each set $F(C)$ associated with the clause C is defined by:

$$F(C) = \{p' : p \text{ appears in } C\} \cup \{p : \neg p \text{ appears in } C\}.$$

From this it follows that the given instance of the satisfiability problem has a solution if and only if the solution of the constructed instance of the CAP is exactly the cardinality of P. First, we present a solution of the CAP based on homogeneous AHNEPs.

Theorem 12. *Let $(S = \{s_1, s_2, ..., s_n\}, F = \{F_1, F_2, ..., F_m\})$, be an (n, m) - instance of the CAP. It can be solved by a complete homogeneous AHNEP of size $m+2n+2$ in $O(m+n)$ time.*

Proof. We construct the complete homogeneous AHNEP:

$$\Gamma = (U, K_{m+2n+2}, N, \alpha, \beta, x_0).$$

Since the result will be collected in a way specified later, the output node is missing. The alphabet of the network is:

$$U = S \cup \hat{S} \cup S' \cup \{Y, Y_1, Y_2, ..., Y_{m+1}\} \cup \{b\} \cup \{Z_0, Z, ..., Z\} \cup \{Y'_1, Y'_2, ..., Y'_{m+1}\} \cup \{X_1, X_2, ..., X_n\},$$

where $\hat{S}$ and S' are copies of S obtained by taking the hat and primed copies of all letters from S, respectively. The nodes of the underlying graph are:

$$x_0, x_F^{(1)}, x_F^{(2)}, \ldots, x_F^{(m)}, x_s^{(1)}, x_s^{(2)}, \ldots, x_s^{(n)}, y_0, y_1, \ldots, y_n.$$

The mapping N is defined by:

$$N(x_0) = (\{X_i \to s_i, X_i \to \hat{s}_i : 1 \le i \le n\} \cup \{Y \to Y_1\} \cup$$
$$\{Y'_i \to Y_{i+1} : 1 \le i \le m\}, \{Y'_i : 1 \le i \le m\}, \varnothing, \varnothing, \{X_i : 1 \le i \le m\} \cup \{Y\}),$$
$$N(x_F^{(i)}) = (\{\hat{s} \to s' : s \in F_i\}, \{Y_i\}, \varnothing, \varnothing, \varnothing), \text{ for all } 1 \le i \le m,$$
$$N(x_s^{(j)}) = (\{s'_j \to \hat{s}_j\} \cup \{Y_i \to Y'_i : 1 \le i \le m\}, \{s'_j\}, \varnothing, \varnothing, \{s'_j\} \cup$$
$$\{Y_i : 1 \le i \le m\}), \text{ for all } 1 \le j \le n,$$
$$N(y_n) = (\{\hat{s}_i \to b : 1 \le i \le n\} \cup \{Y_{m+1} \to Z_0\}, \{Y_{m+1}\}, \varnothing, \{Z_0, b\}, \hat{S}),$$
$$N(y_{n-i}) = (\{b \to Z_i\}, \{Z_{i-1}\}, \varnothing, \{b, Z_i\}, \varnothing), \text{ for all } 1 \le i \le n.$$

The input word is $\{X_1 X_2 \ldots X_n Y\}$. Finally, $\alpha(x) = *$ and $\beta(x) = (1)$, for any node x.

A few words on how the AHNEP works: in the first $2n$ steps, in the first node one obtains 2^n different words $w = x_1 x_2 \ldots x_n Y$, where each x_i is either s_i or $\hat{s}_i$. Each such word w can be viewed as encoding a subset of S, namely the set containing all symbols of S that appear in w. After replacing Y by Y_1 in all of these words, they are sent out and $x_F^{(1)}$ is the only node that can receive them. After one rewriting step, only those words encoding subsets of S that do not include F_1 will remain in the network, the others being lost. The words that remain are easily recognized since they have been obtained by replacing a hat copy of a symbol with a primed copy of the same symbol. This means that this symbol is not in the subset encoded by the word but in F_1. In the nodes $x_s^{(i)}$ the modified hat symbols are restored and the symbol Y'_1 is substituted for Y_1. Now, the words go to the node x_0 where Y_2 is substituted for Y'_1 and the whole process resumes for F_2. This process lasts for $8m$ steps.

The last phase of the computation makes use of the nodes y_j, $0 \le j \le n$. The number we are looking for is given by the largest number of symbols from S in the words from y_n. It is easy to note that the words that cannot leave y_{n-i} have exactly $n-i$ such symbols, $0 \le i \le n$. Indeed, only the words that contain at least one occurrence of b can leave y_n and reach y_{n-1}. Those words that do not contain any occurrence of b have exactly n symbols from S. In y_{n-1}, Z_1 is substituted for an occurrence of b and those words that still contain b leave this node for y_{n-2} and so forth. The words that remain here contain $n-1$ symbols from S. Therefore, when the computation is over, the solution of the given instance of the CAP is the largest j such that y_j is nonempty. The last phase is over after at most $4n+1$ steps. By the aforementioned considerations, the total number of steps is at most $8m+4n+3$; hence, the time complexity of solving each instance of the CAP of size (n, m) is $O(m+n)$.

As far as the time and memory resources the HNEP uses, the total number of symbols is $2m+5n+4$ and the total number of rules is:

$$mn+m+5n+2+card(F_1)+card(F_2)+...+card(F_m)\in \Theta\,(mn).$$

The same problem can be solved in a more economic way, regarding especially the number of rules with HNEPs, namely:

Theorem 13. *Any instance of the CAP can be solved by a complete AHNEP of size $m+n+1$ in $O(m+n)$ time.*

Proof. For the same instance of the CAP as in the previous proof, we construct the complete AHNEP:

$$\Gamma = (U,\ K_{m+n+1},\ N,\ \alpha,\ \beta,\ x_0).$$

The alphabet of the network is:

$$U = S\cup\ S'\cup\ \{Y_1,\ Y_2,\ ...,\ Y_{m+1}\}\cup\ \{b\}\cup\ \{Z_0,\ Z_1,\ ...,\ Z_n\}.$$

The other parameters of the network are given in Table 3.

In Table 3, i ranges from 1 to n and j ranges from 1 to m. The input word is $s'_1s'_2... s'_nY_1$.

Table 3.

Node	M	PI	FI	PO	FO	α	β
x_0	$\{s'_i\rightarrow s_i\}\cup\{s'_j\rightarrow T\}$	$\{s'_1\}$	$\varnothing$	$\varnothing$	$\varnothing$	$*$	(1)
$x_F^{(j)}$	$\{Y_j\rightarrow Y_{j+1}\}$	$\{Y_j\}$	F_j	$\varnothing$	$\varnothing$	$*$	(3)
y_n	$\{T\rightarrow Z_0\}$	$\{Y_{m+1}\}$	$\varnothing$	$\{Y'\}$	T	$*$	(1)
y_{n-i}	$\{T\rightarrow Z_i\}$	$\{Z_{i-1}\}$	$\varnothing$	U	T	$*$	(4)

The reasoning is rather similar to that of the previous proof. The only notable difference concerns the phase of selecting all words that do not contain any symbol from any set F_j. This selection is simply accomplished by the way of defining the filters of the nodes $x_F^{(j)}$. The time complexity is now $2m+4n+1 \in O(m+n)$, while the needed resources are $m+3n+3$ symbols and $m+3n+1$ rules.

CONCLUSIONS AND PERSPECTIVES

We have considered a mechanism inspired from cell biology, namely generating and accepting hybrid networks of evolutionary processors. The nodes of these networks are very simple processors able to perform just one type of point mutation (insertion, deletion, or substitution of a symbol). These nodes are endowed with a filter, which is defined by some random context conditions that seem to be close to the possibilities of biological implementation. A rather suggestive view of these networks is that of a group of connected cells that are similar to each other and have the same purpose, or a *tissue*. However, these very simple mechanisms are computationally complete and able to solve hard problems in polynomial time.

We presented linear solutions for a series of NP-complete problems, namely the three-colorability problem, the bounded Post correspondence problem, and the common algorithmic problem. What other intractable problems could be solved in linear time in this framework is another point of interest to be further investigated. It appears of interest to find the class of these problems. A drawback of our solutions, which is actually common to all DNA-based solutions, is that for any instance of a problem we have to construct an AHNEP. Moreover, the input word may always be considered the empty word; hence, it does not represent an encoding of the given instance. It would be more appropriate to have an AHNEP with a unique underlying structure, consisting possibly of two parts such that the nodes (rules and filters) of one part are fixed while the other nodes are changed in accordance with the given instance. Further, each computation starts on a word encoding the given instance. This is a research direction that we are working on. A natural question regards the free GHNEPs/AHNEPs: What is the computational power of these mechanisms? Are there interesting problems that can be solved by free AHNEPs?

If the filters are defined by membership conditions such as those in Castellanos et al. (2003), then GHNEPs with at most six nodes having filters defined by the membership to a regular language condition are able to generate all recursively enumerable languages no matter the underlying structure. This result is not surprising since similar characterizations have been reported in the literature (Csuhaj-Varjú & Mitrana, 2000; Kari et al., 1997; Kari & Thierrin, 1996; Martín-Vide et al., 1998).

It is worth mentioning some similarities with the membrane systems defined by Păun (2000). In that work, the underlying structure is a tree and the biological data are transferred from one region to another by means of some rules. The main difference is that the rule indicates the target region and the data are sent to and received by that region without any filtering process. A more closely related protocol of transferring data among regions in a membrane system was considered by Martín-Vide et al. (2001).

Although we were not concerned here with a practical implementation, we briefly discuss below an (im)possible real-life implementation of the formal solution of the three-colorability problem proposed in Theorem 10. This implementation is not of biological inspiration as one may expect according to the roots of AHNEPs, but uses the facilities of the World Wide Web. We place in any node of the Web a person who is writing e-mail messages. All the persons in the network know each other (they are friends) and each of them has the e-mail addresses of all of the others stored in his or her address book. For the given instance of the three-colorability problem considered in Theorem 10, the network is formed by $7m+2$ persons, each person corresponding to a node in the complete underlying graph constructed in the proof of Theorem 10. Each person has configured his or her SPAM filter such that incoming messages are rejected if they do not satisfy the random-context conditions determined by the forbidding symbols in the sets FI. This is an easy task since it requires simply checking the absence of some symbols from a finite set in the incoming messages. By using a bit more complicated procedure, each person can also configure his or her e-mail software such that an outgoing message cannot be sent unless it satisfies the random-context conditions defined by the sets FO.

Now the working protocol is that described in the proof of Theorem 10. Every receiver waits for a predefined period of time until the messages sent by his or her friends have been collected in his or her incoming box. This period is settled such that it suffices for all persons. Then each person works in parallel with his or her new incoming messages. Each message is modified in accordance with the substitution rules defined by the sets M. This phase lasts again for a predefined period of time, but long enough that all persons can finish their tasks. Note that this phase can also be carried out by noncomplicated software. Each of the users then simultaneously sends the modified messages to the others and the process resumes. After a while this process halts since, at the same time, nobody excepting a designated person (corresponding to the node x_0) receives any new messages. The incoming box of this person contains all solutions to the given instance of the three-colorability problem. A serious problem is represented by the time needed for modifying the new messages as well as the possibility of storing these messages since the number of new messages received by each person is doubled in any step.

We finish with a natural question: We are conscious that our mechanisms likely have no biological relevance. Then why study them? We believe that by combining our knowledge about behavior of cell populations with advanced formal theories from computer science, we could try to define computational models based on the interacting molecular entities. To this aim we need to accomplish the following:

1.   Understanding which features of the behavior of molecular entities forming a biological system can be used for designing computing networks with an underlying structure inspired by some biological systems.
2.   Understanding how to control the data navigating in the networks via precise protocols.
3.   Understanding how to effectively design the networks.

The results reported in this chapter suggest that these mechanisms might be a reasonable example of global computing due to the real and massive parallelism involved in molecular interactions. Therefore, in our opinion, they deserve a deep theoretical investigation as well as an investigation of biological limits of implementation.

REFERENCES

Ardelean, I., Gheorghe, M., Martín-Vide, C., & Mitrana, V. (2003). A computational model of cell differentiation. In Pre-Proceedings of the *Fifth International Workshop on Information Processing in Cells and Tissues*. Lausanne, 275-287.

Castellanos, J., Leupold, P., & Mitrana, V. (2004). Descriptional and computational complexity aspects of hybrid networks of evolutionary processors. *Theoretical Computer Science*. (in press).

Castellanos, J., Martín-Vide, C., Mitrana, V., & Sempere, J. (2001). Solving NP-complete problems with networks of evolutionary processors. In

Mira, J., & Prieto, A. (Eds.), *IWANN. Lecture Notes in Computer Science*, 2048, 621-628.

Castellanos, J., Martín-Vide, C., Mitrana, V., & Sempere, J. (2003). Networks of evolutionary processors. *Acta Informatica*, 39, 517-529.

Constable, R., Hunt, H., & Sahni, S. (1974). On the computational complexity of scheme equivalence. *Technical Report No. 74-201*, Ithaca, NY: Dept. of Computer Science, Cornell University.

Csuhaj-Varjú, E. & Mitrana, V. (2000). Evolutionary systems: A language generating device inspired by evolving communities of cells. *Acta Informatica*, 36, 913-926.

Csuhaj-Varjú, E. & Salomaa, A. (1997). Networks of parallel language processors. In Păun, Gh., & Salomaa, A. (Eds.), *New trends in formal languages, Lecture Notes in Computer Science*, 1218, 299-318.

Csuhaj-Varjú, E. & Salomaa, A. (2003). Networks of Watson-Crick D0L systems. In Ito, M. & Imaoka, T. (Eds.), *Proceedings of the International Conference Words, Languages and Combinatorics III*, Singapore: World Scientific, 134-150.

Csuhaj-Varjú, E., Dassow, J., Kelemen, J., & Păun, Gh. (1993). *Grammar Systems*. London: Gordon and Breach.

Csuhaj-Varjú, E., Martín-Vide, C., & Mitrana, V. (2003). Hybrid networks of evolutionary processors: Completeness results. (submitted)

Dassow, J., Mitrana, V., & Salomaa, A. (2001). Operations and grammars suggested by the genome evolution. *Theoretical Computer Science*, 270, 1-2, 701-738.

Ehrenfeucht, A., Prescott, D., & Rozenberg, G. (2001). Computational aspects of gene (un)scrambling in ciliates. In Landweber, L., & Winfree, E. (Eds.), *Evolution as computation*, Berlin: Springer, 45-86.

Errico, L., & Jesshope, C. (1994). Towards a new architecture for symbolic processing. In Plander, I. (Ed.) *Artificial Intelligence and Information-Control Systems of Robots '94*, Singapore: World Scientific, 31-40.

Fahlman, S.E., Hinton, G.E., & Seijnowski, T.J. (1983). Massively parallel architectures for AI: NETL, THISTLE and Boltzmann machines. *Proceedings of the AAAI National Conf. on AI*, William Kaufman, Los Altos, CA, 109-113.

Garey, M. & Johnson, D. (1979). *Computers and intractability: A guide to the theory of NP-completeness*. San Francisco, CA: Freeman.

Geffert, V. (1991). Normal forms for phrase-structure grammars. *RAIRO/ Theoretical Informatics and Applications*, 25(5), 473-496.

Gheorghe, M. & Mitrana, V. (2004). A formal language-based approach in biology. *Comparative and Functional Genomics*, 5, 91-94.

Hartmanis, J., Lewis II, P.M., & Stearns, R.E. (1965). Hierarchies of memory limited computations. *Proceedings of the Sixth Annual IEEE Symposium on Switching Circuit Theory and Logical Design*, 179-190.

Hartmanis, J. & Stearns, R.E. (1965). On the computational complexity of algorithms. *Transactions of the American Mathematical Society*, 117, 533-546.

Head, T. (1987). Formal language theory and DNA: An analysis of the generative capacity of specific recombinant behaviours. *Bulletin of Mathematical Biology*, 49, 737-759.

Head, T., Yamamura, M., & Gal, S. (1999). Aqueous computing: Writing on molecules. *Proceedings of the Congress on Evolutionary Computation*, IEEE Service Center, Piscataway, NJ, 1006-1010.

Hillis, W.D. (1985). *The connection machine*. Cambridge, MA: MIT Press.

Kari, L., & Landweber, L. (2000). Computational power of gene rearrangements. *American Mathematical Society*, 54, 207-216.

Kari, L. & Thierrin, G. (1996). Contextual insertion/deletion and computability. *Information and Computation*, *131* (1), 47-61.

Kari, L., Gloor, G., & Yu, S. (2000). Using DNA to solve the bounded correspondence problem. *Theoretical Computer Science*, 231, 193-203.

Kari, L., Păun, Gh. Thierrin, G., & Yu, S. (1997). At the crossroads of DNA computing and formal languages: Characterizing RE using insertion-deletion systems. *Proceedings of the Third DIMACS Workshop on DNA Based Computing*, Philadelphia, 318-333.

Margenstern, M., Mitrana, V., & Pérez-Jiménez, M. (2004). Accepting hybrid networks of evolutionary processors. Submitted.

Martín-Vide, C., Mitrana, V., & Păun, Gh. (2001). On the power of valuations in P systems. *Computación y Sistemas*, 5, 120-128.

Martín-Vide, C., Mitrana, V., Pérez-Jiménez, M., & Sancho-Caparrini, F. (2003). Hybrid networks of evolutionary processors. *Proceedings of GECCO, Lecture Notes in Computer Science*, 2723, Berlin: Springer, 401-412.

Martín-Vide, C., Păun, Gh., & Salomaa, A. (1998). Characterizations of recursively enumerable languages by means of insertion grammars. *Theoretical Computer Science*, *205* (1-2), 195-205.

Ogden, W. (1968). A helpful result for proving inherent ambiguity. *Mathematical Systems Theory*, 2, 191-194.

Păun, Gh. (2000). Computing with membranes. *Journal of Computer and System Sciences, 61*(1), 108-143.

Păun, Gh. (2002). *Membrane computing: An introduction*. Berlin: Springer.

Păun, Gh. & Săntean, L. (1989). Parallel communicating grammar systems: The regular case. *Annals of University of Bucharest, Ser. Matematica-Informatica*, 38, 55-63.

Păun, Gh., Rozenberg, G., & Salomaa, A. (1998). *DNA computing: New computing paradigms*. Berlin: Springer.

Rozenberg, G. & Salomaa, A. (1997). *Handbook of formal languages, vol. 1-3*. Berlin: Springer.

Sankoff, D., Leduc, G., Antoine, N., Paquin, B., Lang, B.F., & Cedergren, R. (1992). Gene order comparisons for phylogenetic inference: Evolution of the mitochondrial genome. *Proceedings of the National Academy of Sciences, USA*, 89, 6575-6579.

Searls, D.B. (1988). Representing genetic information with formal grammars. American Association of Artificial Intelligence. *Proceedings of the 7th National Conference on Artificial Intelligence*, 386-391.

Searls, D.B. (1993). The computational linguistics of biological sequences. In Hunter, L. (Ed.), *Artificial intelligence and molecular biology*. Cambridge, MA: AAAI Press, The MIT Press, 47-120.

Searls, D.B. (1995). Formal grammars for intermolecular structure. In *IEEE Symposium on Intelligence in Neural and Biological Systems*. IEEE Computer Society Press, 30-37.

Searls, D.B. (1997). Linguistic approaches to biological sequences. *Bioinformatics*, 13, 333-344.

Searls, D.B. (2002). The language of genes. *Nature*, 420, 211-217.

Stiebe, R. (2003). Personal communication.

Stiebe, R. (2004). Size considerations for HNEPs. Manuscript.

Yokomori, T. & Kobayashi, S. (1995). DNA evolutionary linguistics and RNA structure modelling: A computational approach. In *IEEE Symposium on Intelligence in Neural and Biological Systems*. IEEE Computer Society Press, 38-45.

Chapter V

Cellular Solutions to Some Numerical NP-Complete Problems:
A Prolog Implementation

Andrés Cordón-Franco, University of Sevilla, Spain

Miguel Angel Gutiérrez-Naranjo, University of Sevilla, Spain

Mario J. Pérez-Jiménez, University of Sevilla, Spain

Agustín Riscos-Núñez, University of Sevilla, Spain

ABSTRACT

This chapter is devoted to the study of numerical NP-complete problems in the framework of cellular systems with membranes, also called P *systems (Păun, 1998). The chapter presents efficient solutions to the subset sum and the knapsack problems. These solutions are obtained via families of P systems with the capability of generating an exponential working space in polynomial time. A simulation tool for P systems, written in Prolog, is also described. As an illustration of the use of this tool, the chapter includes a session in the Prolog simulator implementing an algorithm to solve one of the above problems.*

INTRODUCTION

The race to miniaturize silicon microchips to get more and more powerful (smaller and faster) processors is expected to hit its own physical limits very soon. This is why it is necessary to look for new *unconventional models* of computation. One of the main research lines in this direction is focusing on obtaining new computational paradigms inspired from various well-established natural phenomena in physics, chemistry, and biology. This approach is generically known as natural computing.

This chapter is part of the framework of one of these nature-inspired models, namely, cellular computing with membranes. This model abstracts from the structure and the functioning of a living cell. At the moment it is just at the theoretical level, and it is not likely that it would be implemented in vivo in the near future. However, some simulations *in silico* (i.e., software implementations) have been recently presented, written in various programming languages (Java, C, Scheme, etc.). Although they are not able to actually implement the massive parallelism inherent to the original model, these approaches may be regarded as a proof of concept for this new computational paradigm in dealing with hard problems and as a tool that is able to support both research and pedagogical purposes.

The simulator presented here is written in Prolog, and it was created with the aim of assisting in theoretical research in cellular computing. That is, it is not intended to get an efficient implementation, but to be an intuitive tool that provides faithful and detailed information about the computations taking place within cellular systems. More interestingly, during the development of this tool simulator, we realised that we needed new information that helped the formal verification process of cellular computing.

From Nature to Membrane Computing

In recent years the research field generically named natural computing has been under enormous scrutiny and development. This discipline has started off the investigation of both mathematical models and technological requirements for the implementation of bio-inspired computing paradigms. The research within this field studies the way nature *computes*, conceiving and abstracting new paradigms and computing models.

There are several areas within natural computing that are now well established. *Genetic algorithms* (or, more generally, *evolutionary computing*), introduced by J. Holland (1975), uses some operations inspired by natural evolution and selection in order to improve the process of finding a good solution in a huge set of feasible candidate solutions. *Neural networks,* introduced by W.S. McCulloch and W. Pitts (1943), were inspired by the interconnections and the functioning of neurons in the brain. *Molecular computing* is a research area concerned with the use of molecules as biological hardware to perform computations. *DNA-based* molecular computing was born when L. Adleman (1994), published a solution to an instance of the Hamiltonian path problem by manipulating DNA strands in a lab.

We should mention here *splicing systems,* a notion introduced by T. Head (1987), which constituted the theoretical precursor of this type of computation. This model is not oriented toward performing computations; it is just a formalization of the DNA strand recombinations via restriction enzymes. DNA computing is a subarea of molecular computing that uses DNA strands to take advantage of the huge parallelism provided by the biochemical reactions occurring in a DNA solution. *Membrane computing* was introduced by Păun (1998), and it is inspired from the structure and the functioning of molecules and cells as living organisms able to process and generate information. Indeed, the cells contain different vesicles, each of them delimited by membranes leading to a hierarchical structure. Inside of these vesicles some chemical reactions involving biochemical substances take place, modifying the substances contained in them, but also generating a flow of biochemical elements among different compartments that compose the cell. These processes at the cellular level can be interpreted as computing processes.

When designing a formal system that abstracts the structure and functioning of a cell, there are two ways to follow one can describe, in as much detail as possible, the processes that take place, with the aim of getting a deeper understanding about cells; or one can extract the main characteristics that define a cell, with the intention of obtaining a new computing model — simple but powerful — that allows solving problems that are especially hard in other more classical models. This second approach was the one followed by Păun through transition P systems (1998). Since then, a number of variants of P systems have been considered in the literature (see Păun, 2002, for a comprehensive exposition).

The notion of P systems is directly derived from one of the fundamental components of the cell, the biological membrane. It is well-known that all of the internal compartments that form a cell (even the cell itself) are delimited by membranes. Nevertheless, these membranes do not generate tight compartments, but they allow biological substances to pass through them, most of the time in a selective manner. It is inside these compartments that chemical reactions take place.

Basically, a P system consists of a set of membranes, usually organized in a hierarchical structure (Figure 1). There exists a *skin membrane*, which embraces all the others, separating the system from the external environment. The membranes that do not contain other membranes inside are called *elementary membranes*. The *regions* delimited by the membranes (that is, the space bounded by a membrane and the immediately lower membranes, if there are any) can contain multiple copies of certain objects. By means of fixed evolution rules associated with the membranes (or regions), these objects can evolve producing new objects, and can even travel from one region to an adjacent one, crossing the membranes that separate the system's compartments. Transition P systems offer two levels of parallelism: on the one hand, the rules within a membrane are applied simultaneously; on the other hand, these operations are performed in parallel in all of the regions of the system.

Each region can be seen as a computing unit (a processor), having its own data (biological substances) and its local program (given by biochemical reactions). So, the cell can be seen as an *unconventional computing device*.

Observe that a cellular computing system with membranes is not described through a sequence of basic operations capable of being sequenced over an input data in order to obtain a final result. Instead, a P system is a device whose execution, as for Turing machines, modifies the content of the distinct components that form it until reaching, if so, a halting state, when the system halts. In that sense, this execution could be called *user-independent* since, once the system is constructed, it is not necessary, in principle, to guide it.

Figure 1. Structure of a P system

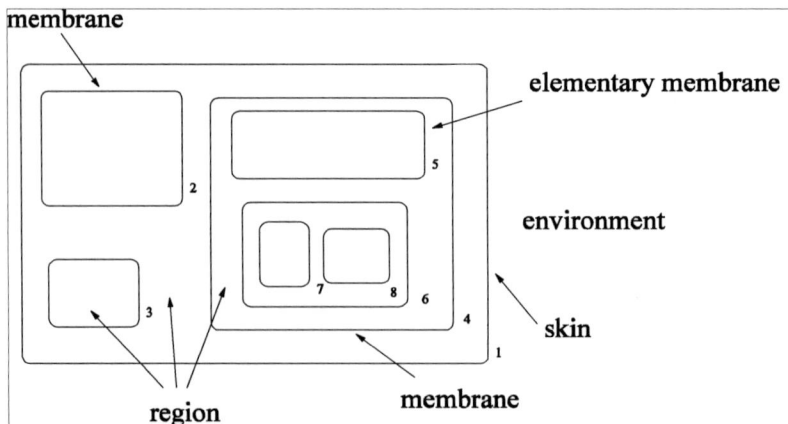

Prolog

The evolution of a P system has a lot of similarities with the execution of a production system based on rules and, because of that, it seems natural to consider a declarative language to simulate cellular computations. In this work we use Prolog (PROgrammation en LOGique) to design a tool for representing and experimenting with P systems in an effective way.

Prolog is a programming language based on clausal logic together with a mechanism of theorem proving (clause resolution). A Prolog program can be regarded as a set of first-order sentences expressing *facts* and *rules*. Providing a comprehensive description of the Prolog language is, of course, beyond the scope of this work (a good starting point can be Bratko, 2001, or *www.afm.sbu.ac.uk/logic-prog/*).

Organization of the Chapter

The chapter is organized as follows: the following section is devoted to briefly describing the class of P systems used in this chapter. In the next section, two families of P systems with active membranes that solve *in linear time* the subset sum and knapsack problems, respectively, are presented and a detailed overview of the computation is given.

The second part of the chapter is devoted to the simulation of the P systems. One section is dedicated to our simulator for P systems with active membranes. After a short presentation about the way of representing P systems in Prolog, the algorithm of the inference engine of the simulator is briefly presented. Finally, we include a subsection showing a Prolog session of the simulator performing the evolution of an instance of the knapsack problem. Interested readers can find more details about the way the simulator works in Cordón Franco et al. (2004).

At the end of the chapter, some conclusions and final remarks are discussed.

RECOGNIZER P SYSTEMS
WITH ACTIVE MEMBRANES

Membrane division is inspired from cell division, a well-known process in biology. The replication is one of the most important functions of a cell. In ideal circumstances, by division it is possible to obtain 2^n cells in n steps. That is, membrane division is able to produce an exponential working space in a polynomial time. This is actually the key feature of P systems with active membranes (Păun, 2002). This characteristic of the P systems allows a significant speed-up in the computation process. Indeed, Zandron (2001) has shown that if **P≠NP**, then a deterministic P system without membrane division is not able to solve an **NP**-complete problem in polynomial time (moreover, this result was generalized in Pérez Jiménez et al. 2004). This speed-up can be especially relevant when dealing with real world problems, such as optimization algorithms in a factory or algorithms to decrypt an encoded message.

Next, following Păun, we introduce the definition of P systems with active membranes and electrical charges. We consider only 2-division for elementary membranes, and we do not use cooperation or priority among rules (Păun, 2002).

Definition

A *P system with active membranes and electrical charges* is a tuple Π =$(\Gamma, H, \mu, w_1, ..., w_q, R)$ where:

- Γ is a finite alphabet (the working alphabet) whose elements are called objects.
- H is a finite set of labels for membranes.
- μ is a membrane structure of degree q, with labels from H. Membranes have electrical charges (0, $+$ or $-$).
- $w_1, ..., w_q$ are multisets over Γ describing the multisets of objects initially placed in membranes from μ.
- R is a finite set of developmental rules of the following types:

    a) $[_l\, a \rightarrow v]_l^{\,\beta}$, where $a \in \Gamma$, $v \in \Gamma^*$, $\beta \in \{0, +, -\}$ (object evolution rules). Internal rules associated with membranes and depending on the label and the charge of the membranes. An object a can evolve to a string v without modifying the polarity of the membrane.

    b) $[_l\, a\,]_l^{\,\beta} \rightarrow b[_l\,]_l^{\,\gamma}$, where $a, b \in \Gamma$, $\beta, \gamma \in \{0, +, -\}$ (communication rules). An object a can get out of a membrane labelled by l and with electrical charge β, possibly transformed in a new one b and, simultaneously, the polarization of the membrane can be changed to γ; its label remains the same.

    c) $a\,[_l\,]_l^{\,\beta} \rightarrow [_l\, b\,]_l^{\,\gamma}$, where $a, b \in \Gamma$, $\beta, \gamma \in \{0, +, -\}$ (communication rules). An object a can get into a membrane labelled by l and with electrical charge β, possibly transformed in a new one b and, simultaneously, the polarization of the membrane can be changed to γ; its label remains unchanged.

d) $[_l a]_l^\beta \rightarrow b$, where $a, b \in \Gamma$, $\beta \in \{0, +, -\}$, $l \neq skin$ (dissolution rules). An object a in a membrane labelled by l and with electrical charge β, is transformed in a new one b and, simultaneously, in the presence of the object a, the membrane is dissolved.

e) $[_l a]_l^\beta \rightarrow [_l b]_l^\gamma [_l c]_l^\delta$, where $a, b, c \in \Gamma$, $\beta, \gamma, \delta \in \{0, +, -\}$, $l \neq skin$ (2-division rules for elementary membranes). In the presence of an object a, the membrane labelled by l and with electrical charge β, is divided into two membranes, eventually allowing independent transformation for the element a on each one of the resulting membranes (i.e., objects b and c, respectively), and possibly the two membranes produced have different polarizations (i.e., γ and δ, respectively).

Let us observe that the rules of the system are associated with labels (for example, the rule $[_l a \rightarrow v]_l^\beta$ is associated with the label $l \in H$). According to rules of type (e), it follows that there may exist membranes in a system with the same label.

The rules are applied according to the following principles:

1.   Their use, according to the general framework of membrane computing, is in a maximal parallel way. In one step, each object in a membrane can only be used by one rule (nondeterministically chosen when there are several possibilities), but any object that can evolve by a rule of any form must do it (with restrictions).

2.   If a membrane is dissolved, its content (multiset and interior membranes) becomes part of the immediately external one. The skin membrane is never dissolved.

3.   All of the elements that are not involved in any of the rules available remain unchanged.

4.   If a division rule is applied to a membrane and, at the same time, some objects inside that membrane evolve through a rule of type (a), then in the two new copies of the membrane we introduce the result of the evolution. That is, we suppose that first the evolution rules of type (a) are used, changing the objects, and then the division is produced, so that in the two new membranes we introduce copies of the changed objects.

5.   The rules associated with the label l are used for all membranes with this label, whether or not the membrane is an initial one or it was obtained through a division process. At one step, different rules can be applied to

different membranes with the same label, but one membrane can be the subject of *only one* rule of types (b), (c), (d), or (e).

6.   The skin membrane can be electrically charged, but can never divide.

Hence, the rules of type (a) are applied in *parallel*; that is, all objects that can evolve by such rules must do it, while the rules of type (b), (c), (d), and (e) are used *sequentially*, in the sense that *one* membrane can be used by, at most, one rule of these types in each step (time unit).

In this work we deal with decision problems, and therefore for each instance of the problem, we are only interested in a binary answer (*yes* or *no*). In this line, there is an underlying similarity between solving a decision problem and solving a recognition problem for a certain language; deciding if an instance has an affirmative or negative solution is equivalent to deciding if a word belongs to the language or not.

We present in this section an adaptation of P systems as devices that accept languages and decide upon some properties. These are called *recognizer P systems*. This adaptation will allow us to efficiently attack some **NP**-complete problems.

Definition

A *P system with input* is a tuple (Π, Σ, i_Π), where:

- Π is a P system, with working alphabet Γ, with q membranes labelled by $1,..., q$, and initial multisets $w_1, ..., w_q$ associated with them.
- Σ is an (input) alphabet strictly contained in Γ.
- The initial multisets are over $\Gamma-\Sigma$.
- i_Π is the label of a distinguished (input) membrane.

Let m be a multiset over Σ. The initial configuration of (Π, Σ, i_Π) with input m is $(\mu, w_1, ..., w_{i\Pi} \cup m, ..., w_q)$. We have thus defined a class of P systems that receive an input before starting the computations. As we intend to consider P systems as black boxes, where we introduce an input and wait for the answer without knowing about the inner processes, we can agree that the output of the computation will be collected outside the P system (through the application of rules of type (b) in the skin membrane). This leads to P systems with input and with external output, which is the variant in which our recognizer systems are included.

Definition

A *recognizer P system with active membranes* is a P system with active membranes, with input and with external output, (Π, Σ, i_{Π}), such that:

- The working alphabet contains two distinguished elements: *yes, no*.
- All of its computations halt.
- If C is a computation of Π, then either an object *yes* or an object *no* (but not both) has to be sent out to the external environment, and only in the last step of the computation.

Let Π be a recognizer P system, and let C be a computation of Π. Let w be the multiset of objects that have been sent out during the computation C. We define then the function *Output* as $Output(C) = yes$ if $yes \in w$ and $Output(C) = no$ otherwise.

We say that C is an *accepting* computation (respectively, *rejecting* computation) if the object *yes* (respectively, *no*) appears in the external environment associated with the corresponding halting configuration of C — that is, if $yes = Output(C)$ (respectively, $no = Output(C)$).

Let us note that a recognizer P system is a *confluent* system in the following sense: every computation with the same initial configuration has the same output.

LINEAR SOLUTIONS TO NP-COMPLETE PROBLEMS

Most of the P systems from the current literature address questions from a formal language theory angle, mainly concerned with universality results and solutions to classical **NP**-complete problems such as SAT or validity (Păun, 2002). We are interested not only in solving new problems, but also in doing so in an *efficient* way. This is why we have chosen P systems with active membranes because, as mentioned in the previous section, it is a model that allows fast solutions to hard problems. We shall focus on the following numerical NP-complete decision problems:

- *Subset sum problem.* Given a finite set, A, a weight function, $w: A \rightarrow \mathbf{N}$, and a constant $k \in \mathbf{N}$, determine whether or not there exists a subset $B \subseteq A$ such that $w(B) = k$.
  It is clear that we can assume that $k \leq w(A)$; otherwise, the solution is always negative.

- *Decision knapsack problem (0/1 bounded version).* Given a knapsack of capacity $k \in \mathbf{N}$, a finite set, A, a weight function, $w: A \rightarrow \mathbf{N}$, a value function, $v: A \rightarrow \mathbf{N}$, and a constant, $c \in \mathbf{N}$, decide whether or not there exists a subset of A such that its weight does not exceed k and its value is greater or equal than c.

We shall use tuples $(n, (w_1, \dots, w_n), k)$ and $(n, (w_1, \dots, w_n), (v_1, \dots, v_n), k, c)$ to represent the instances of the subset sum problem and the knapsack problem, respectively, where n stands for the size of $A = \{a_1, \dots, a_n\}$, $w_i = w(a_i)$, $v_i = v(a_i)$ and k and c are the constants mentioned above.

We shall solve these problems via brute-force algorithms, using recognizer P systems with active membranes with 2-division but without dissolution rules. The solution to these problems may be broken down into several stages:

- *Generation stage*: a single specific membrane for every subset of A is generated via membrane division.
- *Calculation stage*: in each membrane the functions w for the subset sum problem or w and v for the knapsack one are evaluated over the associated subset. This stage will take place in parallel with the previous one.
- *Checking stage*: in each membrane, it is checked whether or not the conditions of the problem for each function are satisfied ($w(B) = k$ for the subset sum problem or $w(B) \leq k$ and $v(B) \geq c$ for the knapsack one). This stage cannot start in a membrane before the previous ones are over in *that* membrane.
- *Output stage*: when the previous stages have been completed in *all* membranes, the system sends out the answer (*yes* or *no*) to the environment.

We shall introduce two families of recognizer P systems with active membranes using 2-division solving in *linear* time the subset sum problem and the knapsack problem.

The Subset Sum Problem

The P systems presented here have a recursive description with respect to the two parameters n and k. In order to express the family using only one parameter (i.e., $\Pi = \{\Pi(t) : t \in \mathbf{N}\}$), we shall use the polynomial-time computable bijection $\langle n, k \rangle = ((n+k)(n+k+1)/2)+n$.

For each $(n, k) \in \mathbf{N}^2$ we consider the P system $(\Pi (\langle n, k \rangle), \Sigma(n, k), i(n, k))$, where the input alphabet is $\Sigma (n, k) = \{x_1, ..., x_n\}$, and $\Pi (\langle n, k \rangle) = (\Gamma (n, k), \{e, s\}, \mu, m_s, m_e, R)$ is defined as follows:

- Working alphabet: $\Gamma (n, k) = \{e_0, ..., e_n, z_0, ..., z_{2n+2k+2}, q, q_0, ..., q_{2k+1}, x_0, ..., x_n, a_0, a, \bar{a}_0, \bar{a}, yes, no, \#, d_+, d_-\}$.
- Membrane structure: $\mu = [_s [_e]_e]_s$
- Initial multisets: $m_s = z_0 ; m_e = e_0 \bar{a}^k$
- Input membrane: $i(n, k) = e$
- The evolution rules from the set R are listed below, classified into several groups that are briefly commented:

(R1) $[_e e_i]_e^0 \rightarrow [_e q]_e^- [_e e_i]_e^+$, for $i = 0, ..., n$,
$[_e e_i]_e^+ \rightarrow [_e e_{i+1}]_e^0 [_e e_{i+1}]_e^+$, for $i = 0, ..., n-1$

The goal of these rules is to generate one membrane for each subset of A. When an object $e_i (i < n)$ is present in a neutrally charged membrane, we pick the element a_i for its associated subset and divide the membrane. In the new membrane where q appears, no further elements will be added to the subset, but the other new membrane must generate membranes for other possible subsets that are obtained by adding elements of index $i + 1$ or greater.

(R2) $[_e x_0 \rightarrow \bar{a}_0]_e^0 ; [_e x_0 \rightarrow \lambda]_e^+ ; [_e x_i \rightarrow x_{i-1}]_e^+$, for $i = 1, ..., n$.

At the beginning, in the input multiset that is introduced before starting the computation, objects x_j (with $1 \leq j \leq n$) encode the weights of the corresponding elements of A: for each a_j we have w_j copies of x_j. Together with elements added to the subset associated with the membrane, these rules calculate the weight of such a subset.

(R3) $[_e q \rightarrow q_0]_e^- ; [_e \bar{a}_0 \rightarrow a_0]_e^- ; [_e \bar{a} \rightarrow a]_e^-$.

The occurrence of the objects q_0, a_0, and a marks the beginning of the checking stage. The multiplicity of object a_0 encodes the weight of the associated subset, and the constant k is represented by the number of objects a.

(R4) $[_e a_0]_e^- \rightarrow [_e]_e^0 \# ; [_e a]_e^0 \rightarrow [_e]_e^- \#$.

We compare the number of occurrences of objects a and a_0, sending them out of the membrane alternatively and changing the polarity of the membrane each time. The two rules of this group describe this *checking loop*.

(R5) $[_e q_{2j} \rightarrow q_{2j+1}]_e^-$, for $j = 0, ..., k$; $[_e q_{2j+1} \rightarrow q_{2j+2}]_e^0$, for $j = 0, ..., k-1$.

Objects q_i, with $0 \leq i \leq 2k+1$, act as a counter for the checking stage, controlling the number of checking loops that take place.

(R6) $[_e q_{2k+1}]_e^- \rightarrow [_e]_e^0 yes$; $[_e q_{2k+1}]_e^0 \rightarrow [_e]_e^0 \#$; $[_e q_{2j+1}]_e^- \rightarrow [_e]_e^- \#$, for $j = 0, ..., k-1$.

Finally, these rules use the information given by the counter to deal with the different checking results — the same number of objects a_0 and a, objects a_0 in excess, or more a objects than a_0.

(R7) $[_s z_i \rightarrow z_{i+1}]_s^0$, for $i = 0, ..., 2n+2k+1$; $[_s z_{2n+2k+2} \rightarrow d_+ d_-]_s^0$.

There is another counter in the skin membrane that waits for all membranes to finish their checking stage and then releases objects d_+ and d_- in the skin.

(R8) $[_s d_+]_s^0 \rightarrow [_s]_s^+ d_+$; $[_s yes]_s^+ \rightarrow [_s]_s^0 yes$; $[_s d_- \rightarrow no]_s^+$; $[_s no]_s^+ \rightarrow [_s]_s^0 no$.

The answering process is now activated: first the object d_+ acts as a query, changing the polarity of the skin membrane, and then any possible object *yes* that may be present in the membrane is sent out (notice that there is no conflict because in this moment there are no objects *no* present in the skin, since the rule $d_- \rightarrow no$ needs a positive charge to be applied).

Let us recall that the evolution rules of $\Pi(\langle n, k \rangle)$ are defined in a recursive manner from the instance u. Furthermore, the necessary resources to build $\Pi(\langle n, k \rangle)$ from a given instance $u = (n, (w_1, ..., w_n), k)$ are the following:

- Size of the alphabet: $4n+4k+17 \in \Theta (n+k)$
- Number of membranes: $2 \in \Theta (1)$
- $|m_s|+|m_e| = k+2 \in \Theta (k)$
- Sum of the lengths of the rules: $35n+27k+110 \in \Theta (n+k)$

At this point, we would like to remark that the values of the main parameters that determine the size of an instance (n and k) are encoded in the system in a unary fashion. Indeed, n is the number of different objects e_i $(i \neq 0)$ that belong to the alphabet and k is the number of copies of $\bar{a}$ that are present in the inner membrane at the beginning of the computation.

Please note that the weights of the elements of A are also introduced in an unary fashion, through the input multiset. However, these weights do not influence the amount of resources needed to build the system, nor the upper bound for the number of steps of any computation. Indeed, let us (informally) calculate this bound.

There is no synchronization among membranes for the generation stage. This stage starts in the first step of the computation for the unique inner membrane of the system, and then the new membranes created by division will evolve independently. However, it can be proved that after $2n+1$ steps no more divisions will take place in the system. The last membrane to leave the generation stage is the one whose associated subset is $B = A$. It is clear that this membrane is also the last one to finish the calculation stage (in the same step). After an additional step is performed, for renaming purposes, the membrane associated with A appears.

Next, in order to complete the third stage, and regardless of its associated subset, each membrane has to perform at most $2k+2$ steps (see rules (R4), (R5), and (R6)). The exact number of steps is less if $w(B) < k$, but we are looking for an upper bound. Therefore, after $2n+2k+4$ steps all of the inner membranes have completed the first three stages.

Now let us focus on the skin membrane for the last stage. The counter z_j is working from the very beginning of the computation, and in the $(2n+2k+3)$-th step it evolves into the objects d_- and d_+. The latter object then leaves the system, preparing the skin to send out the final answer (by changing the charge of the skin to positive). In this moment no rules are applicable in any inner membrane, and the only ones that can be applied are the ones from (R8). Thus, the total number of steps will be $2n+2k+5$, if the answer is affirmative, or $2n+2k+6$, otherwise. In this way we can say that the family $\Pi = (\Pi\,(t))_{t \in \mathbf{N}}$ solves the subset sum problem in *linear* time.

Notice that the answer will be sound because all of the checking stages for all of the membranes have been completed (and, thus, all of the subsets have been tested) before checking the presence of an object *yes* in the skin membrane by rules (R8).

The Knapsack Problem (0/1 Bounded Version)

As we did in the previous subsection for the subset sum problem, let us present now a family of P systems that solves the knapsack problem. We shall not again discuss the groups of rules; the reader is encouraged to get through the list of rules and to find out how the computation develops in this case.

Now the relevant parameters for the design of the P systems are n, k, and c. We shall consider a computable polynomial bijection between $\mathbf{N}^3$ and $\mathbf{N}$ (e.g., the one induced by the pairing function $\langle x, y, z \rangle = \langle x, \langle y, z \rangle \rangle$).

The family presented here is $\Pi = \{(\Pi(\langle n, k, c \rangle), \Sigma(n, k, c), i(n, k, c)) : (n, k, c) \in \mathbf{N}^3\}$, where the input alphabet is $\Sigma(\langle nkc \rangle) = \{x_1, ..., x_n, y_1, ..., y_n\}$, and the P system $\Pi(\langle n, k, c \rangle) = (\Gamma(n, k, c), \{e, s\}, \mu, m_s, m_e, R)$ is defined as follows:

- Working alphabet: $\Gamma(n, k, c) = \{a_0, a, \bar{a}_0, \bar{a}, b_0, b, \underline{b}_0, \underline{b}, \underline{\underline{b}}_0, \underline{\underline{b}}, d_+, d_-, e_0,$
  $... , e_n, q, q_0, ..., q_{2k+1}, r, r_0, ..., r_{2c+1}, x_0, ..., x_n, y_0, ..., y_n, yes, no, z_0, ...,$
  $z_{2n+2k+2c+6}, \#\}$.
- Membrane structure: $\mu = [_s [_e]_e]_s$.
- Initial multisets: $m_s = z_0$; $m_e = e_0 \bar{a}^k \underline{b}^c$.
- Input membrane: $i(n, k, c) = e$.
- Evolution rules, R, with the following rules:

$[_e e_i]_e^0 \rightarrow [_e q]_e^- [_e e_i]_e^+$, for $i = 0, ..., n$.
$[_e e_i]_e^+ \rightarrow [_e e_{i+1}]_e^0 [_e e_{i+1}]_e^+$, for $i = 0, ..., n-1$.

$[_e x_0 \rightarrow \bar{a}_0]_e^0$; $[_e x_0 \rightarrow \lambda]_e^+$; $[_e x_i \rightarrow x_{i-1}]_e^+$, for $i = 1, ..., n$.
$[_e y_0 \rightarrow \underline{b}_0]_e^0$; $[_e y_0 \rightarrow \lambda]_e^+$; $[_e y_i \rightarrow y_{i-1}]_e^+$, for $i = 1, ..., n$.

$[_e q \rightarrow r q_0]_e^-$; $[_e \bar{a}_0 \rightarrow a_0]_e^-$; $[_e \bar{a} \rightarrow a]_e^-$; $[_e \underline{b}_0 \rightarrow \underline{\underline{b}}_0]_e^-$; $[_e \underline{b} \rightarrow \underline{\underline{b}}]_e^-$.

$[_e a_0]_e^- \rightarrow [_e]_e^0 \#$; $[_e a]_e^0 \rightarrow [_e]_e^- \#$.

$[_e q_{2j} \rightarrow q_{2j+1}]_e^-$, for $j = 0, ..., k$; $[_e q_{2j+1} \rightarrow q_{2j+2}]_e^0$, for $j = 0, ..., k-1$.

$[_e q_{2j+1}]_e^- \rightarrow [_e]_e^+ \#$, for $j = 0, ..., k$.

$[_e r \rightarrow r_0]_e^+$; $[_e \underline{\underline{b}}_0 \rightarrow b_0]_e^+$; $[_e \underline{\underline{b}} \rightarrow b]_e^+$; $[_e a \rightarrow \lambda]_e^+$.

$[_e b_0]_e^+ \rightarrow [_e]_e^0 \#$; $[_e b]_e^0 \rightarrow [_e]_e^+ \#$.

$[_e \, r_{2j} \rightarrow r_{2j+1} \,]_e^+$, for $j = 0,..., c$; $[_e \, r_{2j+1} \rightarrow r_{2j+2} \,]_e^0$, for $j = 0,..., c-1$.

$[_e \, r_{2c+1} \,]_e^+ \rightarrow [_e \,]_e^0 \, yes$; $[_e \, r_{2c+1} \,]_e^0 \rightarrow [_e \,]_e^0 \, yes$.

$[_s \, z_i \rightarrow z_{i+1} \,]_s^0$, for $i = 0,... ,2n+2k+2c+5$; $[_s \, z_{2n+2k+2c+6} \rightarrow d_+ \, d_- \,]_s^0$.

$[_s \, d_+ \,]_s^0 \rightarrow [_s \,]_s^+ \, d_+$; $[_s \, yes \,]_s^+ \rightarrow [_s \,]_s^0 \, yes$; $[_s \, d_- \rightarrow no \,]_s^+$; $[_s \, no \,]_s^+ \rightarrow [_s \,]_s^0 \, no$.

Before continuing, we would like to stress the fact that the evolution rules of $\Pi(\langle n, k, c \rangle)$ are described in a recursive manner from the instance u. Let us also list, as we did in the previous section, the resources needed to build Π $(\langle n, k, c \rangle)$:

- Size of the alphabet: $5n+4k+4c+28 \in \Theta(n+k+c)$
- Number of membranes: $2 \in \Theta(1)$
- $|m_s|+|m_e| = k+c+2 \in \Theta(k+c)$
- Sum of the lengths of the rules: $40n+27k+20c+193 \in \Theta(n+k+c)$

Keeping in mind what was discussed about the unary encoding of the parameters and the number of steps of the computations for the subset sum problem, we say that the family presented in this section solves the knapsack problem in *linear* time.

An Overview of the Computation

This section is devoted to explaining the way the P systems described earlier work to solve numerical problems. As the solutions presented for the subset sum and knapsack problems are very similar, we shall only discuss the first.

First of all, recall that to solve an instance $u = (n, (w_1, ..., w_n), k)$ of the subset sum problem we take the P system $\Pi(\langle n, k \rangle)$ with input $x_1^{w1}...x_n^{wn}$. We shall, therefore, refer to such P systems with these inputs from now on.

The purpose of the first stage (generation) is to get a single *relevant* membrane for each subset of A (the concept of relevant membrane is given below). This means 2^n different relevant membranes in all.

In the first step of the computation, the rule $[_e \, e_0 \,]_e^0 \rightarrow [_e \, q \,]_e^- [_e \, e_0 \,]_e^+$ is applied. Then the generation and calculation stages continue on in parallel, following the instructions given for the rules (R1) and (R2). These two stages do not end in a membrane as long as an object e_i (with $0 \le i \le n$) belongs to it and its charge is positive or neutral.

Let us introduce the concept of subset associated with an internal membrane through the following recursive definition:

- The subset associated with the initial inner membrane is the empty one.
- When an object e_j appears in a neutrally charged membrane (with $j < n$), then the j-th element of A is selected and added up to the previous associated subset. Once the stage is over, the associated subset will not be modified anymore.
- When a division rule is applied, the two newborn membranes inherit the associated subset from the original membrane.

We shall also refer to a membrane as associated with its corresponding subset.

After a division rule $[_e\ e_i\ ]_e^0 \rightarrow [_e\ q\ ]_e^- [_e\ e_i\ ]_e^+$ is applied, the two new membranes will behave in a quite different way. On the one hand, in the negatively charged membrane (we have marked such membranes in Figure 2 with a circle), the two first stages end, and in the next step the rules in the group (R3) are applied, renaming the objects to prepare for the third stage. This step is a significant moment, so we shall call *relevant* membranes those that have a negative charge and contain an object q_0. A relevant membrane will not divide anymore during the computation, and its associated subset will remain unchanged. On the other hand, the positively charged membrane will continue in the generation stage. This stage will give rise to membranes associated with subsets that are obtained by adding elements of index $i+1$ or greater to the current one. Note that if $i = n$, then the membrane cannot continue the generation stage, as there are no rules working for an object e_n in a positively charged membrane (see the membranes surrounded by a diamond in Figure 2). It makes sense that these membranes get blocked, as it is not possible to add elements of indices greater than n.

Thus, as the indices of objects e_i never decrease, we notice that the relevant membranes are generated in a sort of "lexicographic order", in the following sense: if the j-th element of A has already been added to the associated subset, then no element with index lower than j will be added later to the subset associated with that membrane nor to the subsets associated with its descendants. We can check that every subset of A will get a single relevant membrane associated with it, but these membranes are not generated altogether simultaneously. Indeed, we can also check that the membrane corresponding to the subset $\{a_{i1}, ..., a_{ir}\}$ will arise in the $(i_r + r + 2)$-th step.

As noted earlier, the first two stages are carried out in parallel. Indeed, there is only a gap of one step of computation between the moment when an element is added to the associated subset and the moment when the new weight of the subset is updated. It can be proved that in a positively charged membrane $[_e]_e^+$ where the object e_i occurs, the multiplicity of object x_1 is equal to the weight of element $a_{i+1} \in A$. Thus, when in the next step we apply (simultaneously) the rules $[_e x_1 \to x_0]_e^+, [_e x_0 \to \lambda]_e^+$ and $[_e e_i]_e \to [_e e_{i+1}]_e^0 [_e e_{i+1}]_e^+$, each child membrane will contain exactly $w(a_{i+1})$ occurrences of x_0.

In the case of the neutrally charged child, the element a_{i+1} is added to the associated subset and, simultaneously, the weight of this subset is updated because all of the objects x_0 mentioned above will be transformed into $w(a_{i+1})$ copies of $\bar{a}_0$. The situation is different for the positively charged child, where the element a_{i+1} will not be added to the associated subset or to any of the subsets associated with the descendant membranes. In this case, we are not interested in the weight of the object a_{i+1}, so all the objects x_0 present in the membrane are removed, while $w(a_{i+2})$ new copies of x_0 are created. This procedure goes on until getting into a relevant membrane, and then the number of occurrences of object a_0 will encode exactly the weight of the associated subset and the membrane will be ready to begin the next stage (checking).

Figure 2. Membrane generation for n = 3

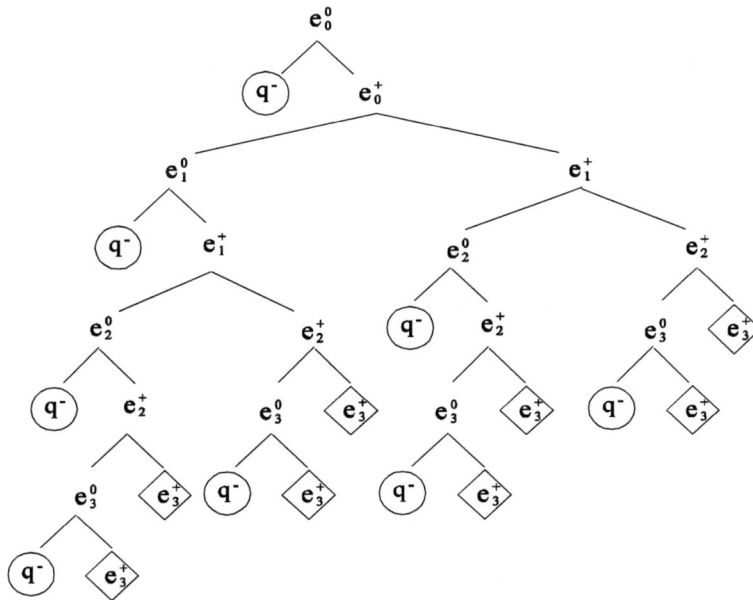

For example, for a_1, after two steps of computation we can see that there are three inner membranes in the configuration, and one of them is a neutrally charged membrane where the object e_1 occurs (see Figure 2). Thus, the element a_1 is added to the associated subset. It can be proven that there are, at that moment, w_1 copies of x_0 in the membrane. So, in the next step, at the same time as the membrane divides, w_1 objects $\bar{a}_0$ will be generated in the membrane by the rules in (R2). These rules will also modify the indices of all the objects x_i, with $i > 0$, so that w_2 copies of x_0 will be ready in the membrane in the next step.

The purpose of the rules in (R4) is to compare the multiplicities of objects a_0 and a — that is, to perform the checking stage for w. They will be counted one by one alternatively, changing the membrane charge each time from negative to neutral and vice versa. In the case of the subset sum problem, the checking is successful if and only if $w(B) = k$, then after $2k$ steps of the checking stage we will not have any objects a or a_0 and the charge will be negative. The counter q_i controls if the number of checking steps is actually equal to $2k$, via the rules in (R6). Let us describe in Table 1 how the third stage works in a membrane associated with a subset B of weight w_B.

Please observe that the index of q_i coincides with the total amount of copies of a and a_0 that have already been erased during the comparison stage.

If $B = \{a_{i1}, ..., a_{ir}\}$ with $i_r \neq n$, then there are in the multiset some objects x_j for $1 \leq j \leq n - i_r$, but they are irrelevant for this stage and, therefore, they are omitted.

If the number w_B of objects a_0 is equal to the number k of objects a, then the result of this stage is successful. If the number of objects a_0 is greater than k, then every time that the rule $[_e q_{2j} \rightarrow q_{2j+1}]_e^-$ is applied (that is, for $j = 0, ..., k$), the rule $[_e a_0]_e^- \rightarrow [_e]_e^0 \#$ will also be applied. Thus, we can never find a situation

Table 1.

Multiset	Charge	Parity of q_i
$q_0\, a_0^{w_B}\, a^k$	–	EVEN
$q_1\, a_0^{w_B-1}\, a^k$	0	ODD
$q_2\, a_0^{w_B-1}\, a^{k-1}$	–	EVEN
...	...	...
$q_{2j}\, a_0^{w_B-j}\, a^{k-j}$	–	EVEN
$q_{2j+1}\, a_0^{w_B-(j+1)}\, a^{k-j}$	0	ODD
...	...	...

in which the index of the counter q_i is an odd number (namely, $i = 2k+1$) and the charge is negative. This means that the rule that sends out an object *yes* to the skin can never be applied, and even more, the membrane gets *blocked* (it will not evolve anymore during the computation).

Finally, rules in (R7) and (R8) are associated with the skin membrane and complete the output stage. The counter z_i waits before releasing the objects d_+ and d_- to avoid the answer being sent out before all of the inner membranes have finished their checking stages (or have been blocked). The generation and calculation stages will last at most $2n+2$ steps; after those steps, one transition step is performed (rules in (R3)), and the checking process for w is bounded by $2k+1$. This makes $2n+2k+2$ steps in all. After all of these steps are performed, the output process is activated.

In that moment, the skin will be neutrally charged and will contain the objects d_+ and d_-. Furthermore, some objects *yes* will be present in the skin if and only if the checking stages have been successful in at least one inner membrane.

The output stage then begins. First, the object d_+ is sent out, giving positive charge to the skin. Then the object d_- evolves to *no* inside the skin and, simultaneously, if any object *yes* is present in the skin, it will be sent out of the system, giving neutral charge to the skin and making the system stop (in particular, further evolution of object *no* is avoided).

If none of the membranes has successfully passed its checking stage, then any object *yes* will not be present in the skin membrane when the output stage begins. Thus, after generating an object *no*, the skin membrane will still have a positive charge and will be sent out. In that moment, the system halts.

A PROLOG IMPLEMENTATION FOR P SYSTEMS WITH ACTIVE MEMBRANES

Choosing a programming language for implementing a model of computation is not an easy decision. It is necessary to analyze the main difficulties to develop such an implementation and look for a programming language with the adequate features to solve them. As far as our model is concerned, formalizing a configuration of a P system involves dealing with complex data structures. We have to both represent the membrane structure of the P system and make explicit the content of every membrane. Hence, the chosen language has to be expressive enough to handle symbolic knowledge in a natural way.

Moreover, the rules of a P system are nearer to a *production* system than to a list of instructions to be executed in a sequential way. Thus, it seems natural to choose a declarative language (the programmer specifies *what* is to be computed) rather than an imperative one (the programmer specifies *how* something is to be computed).

Prolog is expressive enough to handle symbolic knowledge in a quite natural way and has the ability to evolve different configurations following a set of rules in a declarative style. Aside from the based-tree data structure and the use of infix operators defined ad hoc by the programmer, it allows us to imitate natural language, and the user can follow the evolution of the system without any knowledge of Prolog. We refer to the authors' Web page for a detailed description of the simulator (the Prolog files of the simulator, together with a user manual, are freely available by e-mail from the authors and will soon be found at *www.gcn.us.es*).

In the current version, our Prolog simulator for P systems with active membranes consists of two different parts, as shown in Figure 3. The first part is an *inference engine*. This is a Prolog program that takes as input an initial configuration and the set of rules of a P system and carries out the evolution process associated with the system. Let us emphasize that the inference engine is completely general; that is, it does not depend on the considered P system at all. The second part of the simulator (the program *generator.pl*) provides a tool to automatically build the initial configurations and the sets of rules for instances of some well-known **NP**-complete problems (e.g., SAT, validity, subset sum, and knapsack problems).

Our simulator looks in the ordered set of rules for all those that are to be applied, carries out the corresponding evolution step, and deterministically obtains a unique new configuration. Consequently, our simulator only ensures a correct simulation of evolutions of deterministic or confluent P systems. Nevertheless, most of the usual algorithms that solve interesting problems are covered.

In what follows, we describe the two parts of our P system simulator (the *inference engine* and the program *generator.pl*). For the sake of simplicity, we shall skip technical details.

Configuration Representation

The first decision related to the design of the simulator is the way to represent the knowledge domain. The problem of representing a model into a programming language is a universal question in computer science, but in this case it has a double face: on one hand, the choice has to allow an efficient

Figure 3. The Prolog simulator

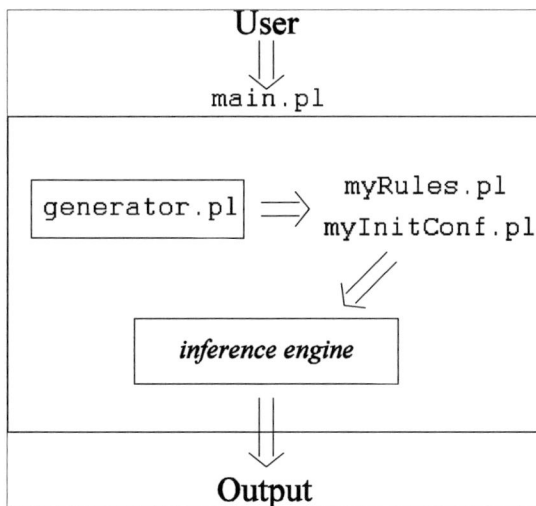

treatment of the data from a programmer's point of view, and on the other hand, the design has to be intuitive enough to be understood by users, regardless of their knowledge of Prolog. These questions lead us to define several infix operators that keep the logical notation of Prolog and allow the reading of the information in English-like sentences. The usual notation for membranes and rules by using subscripts and superscripts has been replaced by usual, plain-text representation in English-like Prolog code.

In order to specify a configuration for a P system with active membranes, it is necessary to explicitly represent the membrane structure and the content of every membrane present in this structure. This involves representing the label, electrical charge, and position of each membrane as well as its content. Moreover, we shall also keep track of the current step of the computation.

In our model, the configuration of a P system in each step of the evolution is a set of one-literal clauses, each of them representing one *alive* membrane (a membrane that is still part of the current system and is not yet dissolved; consequently, it disappears for the next step of computation). Hence, each clause will show a label, an electrical charge, a position, a multiset of objects, and the current step of the computation, as well as the P system to which this membrane belongs.

To achieve this step, the operators ::, **ec**, **at**, **with** and **at_time** are used. Then, to denote that in the **t**-th step of its evolution the P system **P** has a membrane at position [pos] with label h, polarization α, and m as the multiset, we shall write:

P :: h ec α at [pos] with m at_time t

Rules Description

Next we present the general form of the different types of rules of a P system with active membranes (where P is the name of the current P system):

a) $[_l\, a \rightarrow v]_l^\alpha$ (where $v = v_1 \dots v_n$)

P rule a evolves_to [v1,...,vn] in l ec α.

b) $[_l a\,]_l^\alpha \rightarrow b[_l\,]_l^\beta$

P rule a inside_of l ec α sends_out b of l ec β.

c) $a\,[_l]_l^\alpha \rightarrow [_l b\,]_l^\beta$

P rule a out_of l ec α sends_in b of l ec β.

d) $[_l\, a\,]_l^\alpha \rightarrow b$

P rule a inside_of l ec α dissolves_and_sends_out b.

e) $[_l\, a\,]_l^\alpha \rightarrow [_l\, b\,]_l^\beta [_l\, c\,]_l^\gamma$,

P rule a inside_of l ec α divides_into

   b inside_of l ec β and c inside_of l ec γ.

The Algorithm

The Prolog algorithm to simulate the evolution of a P system works in quite naturally. The input of the program is the initial configuration of the membranes, which is represented as a set of sentences (one-literal clauses with predicate symbol ** and all of them at_time 0) and an appropriate set of rules.

Step 1. *Initialization.* At the beginning, all membranes are set to *applicable*, and for each membrane, their objects are split into two multisets: one *usable* multiset, containing all of the objects of the membrane, and one *used* multiset, which is initially empty.

Step 2. *Transition.* If an applicable membrane enabling the associated rules to be applied exists, then these rules are applied in the following way:

(a)-stage. At this stage, only rules of type (a) are checked one by one. If an object triggers the rule, then it is removed from *usable* to prevent one object being used by two different rules at the same step. The multiset resulting from the application of the rule is added to *used*. After this step, the membrane remains applicable and new evolution rules can be applied. This stage ends when no more rules of type (a) can be applied.

Non-(a)-stage. At this stage, only one rule of types (b), (c), (d), or (e) can be applied. Let us remember that this simulation works with deterministic or confluent P systems. The action depends on the type of rule to be applied:

- *Send out rule.* The element that triggers the rule is removed from the usable multiset and the new one is added to the used multiset of the father membrane. The membrane changes to *not applicable* mode. If the element is sent out of the skin, then it is marked with the *outside* label.
- *Send in rule.* This rule is the opposite of the send out rule. The element that triggers the rule is removed from the usable multiset in the father membrane and the new one is added to the used multiset. The membrane changes to not applicable mode.
- *Dissolution rule.* The element that triggers the rule is removed from the usable multiset and the new element obtained, together with the rest of the elements of the membrane, is added to the used multiset of the father membrane. The membranes inside the dissolved membranes become children of the father membrane.
- *Division rule.* The element that triggers the rule is removed from the usable multiset and the division creates two new membranes in not applicable mode. One of them keeps the original position and the second one gets a new position. The current version of the simulator can also deal with 2-division rules of nonelementary membranes, although this feature falls out of the scope of this work.

End step. When no more rules can be applied to membranes in applicable mode, a new configuration (with **at_time** incremented by 1) is stored. At this moment no membrane is into applicable or not applicable mode. These modes only make sense during the evolution step. Now the P system is ready for a new step of the evolution.

Step 3. End of computation. The evolution of the P system finishes when there are no rules to be applied.

A PROLOG SESSION

In this section we present a brief description of how to use the simulator, which consists of several Prolog files that are designed to run on SWI-Prolog 5.2.0, or above, available from *www.swi-prolog.org*.

In the sequel a session for one instance of the knapsack problem is shown. We consider a set $A = \{a_1, a_2, a_3, a_4\}$ ($n = 4$), with weights $w(a_1) = 3$, $w(a_2) = 2$, $w(a_3) = 3$, and $w(a_4) = 1$, and values $v(a_1) = 1$, $v(a_2) = 3$, $v(a_3) = 3$, and $v(a_4) = 2$. The question is to decide whether or not there exists $B \subseteq A$ such that $w(B) \leq 3$ ($k = 3$) and $v(B) \geq 4$ ($c = 4$). According to the presentation above, the P system that solves this instance is Π (‹4, 3, 4›) with input $x_1^3 x_2^2 x_3^3 x_4 y_1 y_2^3 y_3^3 y_4^2$.

The following two Prolog facts allow us to represent this initial configuration (we have chosen the name p1 to denote the P system that solves this instance of the problem). Note that we use ASCII symbols to represent the objects of the alphabet: a_ stands for $\bar{a}$, b0g stands for $\underline{b}$, q_ stands for r, and so on.

```
p1 :: s ec 0 at [] with [z0] at_time 0.
p1 :: e ec 0 at[1] with [e0, a_, a_, a_, b_, b_, b_, b_,
                 x1, x1, x1, x2, x2, x3, x3, x3,
                 x4, y1, y2, y2, y2, y3, y3, y3,
                 y4, y4] at_time 0.
```

The simulator automatically generates these symbols from the data introduced by the user and stores them in a text file. It may be observed that the multiplicities of the objects x_j and y_j correspond to the weights and values of the elements $a_j \in A$, respectively. Also, there are three copies of a_ ($k = 3$) and four copies of b_ ($c = 4$).

The simulator also generates the set of rules associated with the instance of the problem. This generation is completed by instantiating several schemes of rules to the concrete values of the parameters. This produces a text file containing the rules that can easily be edited, modified, or reloaded by the user. The set of rules only depends on the parameters n, k, and c. Consequently, if we want to solve several instances with the same parameters, we only need to generate the set of rules once. In this example, we have obtained 89 rules. Some of them are listed in the appendix at the end of this chapter.

To start with the simulation of the evolution of the P system p1 from the time 0, we enter the following command: evolve(p1,0). The simulator returns the configuration at time 1 and the set of used rules indicating how many times they have been used. Moreover, if the skin sends out any object, then this will be reported to the user. The multisets are represented as lists of pairs obj-n, where obj is an object and n is the multiplicity of obj in the multiset.

```
?- evolve(p1,0).

p1 :: s ec 0 at [] with [z1-1] at_time 1
p1 :: e ec -1 at [1] with  [a_-3, b_-4, q-1, x1-3, x2-2, x3-3,
                            x4-1, y1-1, y2-3, y3-3, y4-2] at_time 1
p1 :: e ec 1 at [2] with [a_-3, b_-4, e0-1, x1-3, x2-2, x3-3,
                            x4-1, y1-1, y2-3, y3-3, y4-2] at_time 1

Used rules in the step 1:
  * The rule 1 has been used only once
  * The rule 57 has been used only once
Yes.
```

In this step only rules 1 and 57 have been applied. Rule 1 is a division rule, the membrane labeled by e at position [1] divides into two membranes that are placed at positions [1] and [2], rule 57 is an evolution rule, and z_0 evolves to z_1 in the skin membrane. To obtain the next configuration in the evolution of p1, now we type:

```
?- evolve(p1,1).

p1 :: s ec 0 at [] with [z2-1] at_time 2
p1 :: e ec -1 at [1] with [a-3, bg-4, q0-1, q_-1, x1-3,
                x2-2, x3-3, x4-1, y1-1, y2-3, y3-3, y4-2] at_time 2
p1 :: e ec 0 at [2] with [a_-3, b_-4, e1-1, x0-3, x1-2,
                x2-3, x3-1, y0-1, y1-3, y2-3, y3-2] at_time 2
```

p1 :: e ec 1 at [3] with [a_-3, b_-4, e1-1, x0-3, x1-2,
              x2-3, x3-1, y0-1, y1-3, y2-3, y3-2] at_time 2

Used rules in the step 2:
  * The rule 6 has been used only once
  * The rule 14 has been used 3 times
  * The rule 15 has been used 2 times

      ...
  * The rule 22 has been used only once
  * The rule 25 has been used 4 times
  * The rule 26 has been used 3 times
  * The rule 58 has been used only once

In this step the first relevant membrane, associated with the empty set, appears at position [1]. The membrane at position [2] will continue dividing to generate new membranes. All of the subsets associated with the membrane's descendant membranes will contain the element a_1. On the other hand, the membrane at position [3] is responsible of the membranes corresponding to all the nonempty subsets that do not contain a_1. Notice that rules from (R2) have been applied (see appendix). That is, the calculation stage has already begun.

The purpose of the first two stages of the system is to generate a single relevant membrane for each subset of A (i.e., $2^n = 2^4 = 16$ relevant membranes in all) and, in parallel, to calculate the weight and the value of such subsets. Other nonrelevant membranes are generated as well, due to technical reasons. These membranes are generated in the first $2n+2 = 10$ steps of the computation.

We can go directly to the configuration at time 10, skipping the intermediate steps, by typing the following command:

?- configuration(p1,10).

      ...
p1 :: e ec -1 at [12] with [a-2, a0-2, b0g-5, bg-4, q2-1, q_-1]
                                      at_time 10
      ...
p1 :: e ec -1 at [24] with [a-3, a0-9, b0g-9, bg-4, q0-1, q_-1]
                                      at_time 10
      ...

Observe that the relevant membrane associated with the total subset appears at position [24]. It is the last relevant membrane to be generated; that is, no more division will take place in the rest of the computation and no new relevant membranes will appear.

Let us focus on the membrane at position [12]. This membrane encodes the subset $\{a_2, a_4\}$, which is the only solution for the instance considered in our example. Two steps of the checking stage for w have already been carried out (note the counter q2). In the next steps the membranes perform their checking stages (for w and for v) that mainly consist of applying rules 27 and 28 (for w) and rules 44 and 45 (for v). Let us focus now on the output stage.

```
?- configuration(p1,26).

p1 :: s ec 0 at [] with [# -127, z26-1] at_time 26
        ...
p1 :: e ec 0 at [12] with [q_9-1] at_time 26
        ...
```

The inner membrane at position [12] is now ready to send an object **yes** to the skin membrane (see rule 56 in the appendix).

```
?- evolve(p1,26).

p1 :: s ec 0 at [] with [# -127, yes-1, z27-1] at_time 27
        ...
p1 :: e ec 0 at [12] with [] at_time 27
        ...
```

Used rules in the step 27:
  * The rule 56 has been used only once
  * The rule 83 has been used only once

Due to technical reasons, the counter in the skin will wait two more steps before releasing the special objects d_+ and d_0.

```
?- configuration(p1,29).

p1 :: s ec 0 at [] with [# -127, d1-1, d2-1, yes-1] at_time 29
```

In this step all inner processes are over. The only object that evolves is z28 (see the set of rules (R11) in the appendix, noting that $28 = 2n+2k+2c+6$).

?- evolve(p1,29).

p1 :: s ec 1 at [] with [# -127, d2-1, yes-1] at_time 30
   ...

Used rules in the step 30:
   * The rule 86 has been used only once
The P-system has sent out d1 at step 30

In this step the object d1 leaves the system and the skin gets a positive charge.

?- evolve(p1,30).

p1 :: s ec 0 at [] with [# -127, no-1] at_time 31
   ...

Used rules in the step 31:
   * The rule 87 has been used only once
   * The rule 88 has been used only once
The P-system has sent out d1 at step 30
The P-system has sent out yes at step 31

In this step, the object **yes** is sent out of the system. To check the system we try to evolve one more time, though this is a halting configuration.

?- evolve(p1,31).
No more evolution!
The P-system p1 has already reached a halting
configuration at step 31

Statistics

Figure 4 shows the evolution along the computation of several indices: the number of membranes, the total number of objects in the P system, and the number of applications of rules that take place. Note that at step nine, the maximum number of membranes is reached, which corresponds to the maxi-

Figure 4. Statistics of the simulation

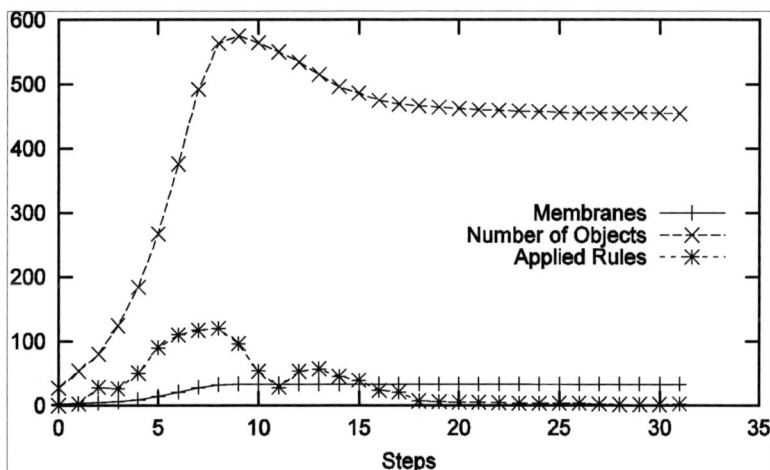

mum number of objects in the P system. Recall that the generation stage ends at time $2n+1 = 9$. From this point, the number of membranes remains constant because no membrane is divided or dissolved, but the number of objects in the P system begins to decrease. Note also that after objects are renamed (calculation stage), the number of applied rules decreases significantly since in the next stage (checking for w) only two rules are applied in each relevant membrane.

In the last steps of the computation, the total number of objects remains almost unchanged because the checking stages are finished in all membranes but one.

The reader is invited to run simulations for other instances of the problem or to test her or his own approach involving P systems with active membranes.

This simulator is not intended to be an efficient software, but a useful assistant that helps the formal verifications of cellular systems. In our example ($n = 4$, $c = 4$, $k = 3$), with an AMD Athlon™ processor at 1.8 GHz and 480 MB RAM, the simulation took 12.23 seconds to perform 8,045,941 Prolog inferences in order to simulate 31 cellular steps.

FINAL REMARKS

The first cellular solutions to **NP**-complete problems using P systems with no input membrane have been described (see Păun, 2001; Zandron, 2001). In such circumstances, specific P systems are built associated with *each* particular instance of the problem to solve. In this chapter, P systems with input are designed, each of them allowing the process of a set of instances of the problem (all those that have the same size, with respect to a predefined criterion).

Although P systems are in general nondeterministic devices, in this chapter we consider recognizer P systems, which provide the same output for all of the computations associated with the same input data. Thus, it is not relevant which branch of the computation is actually chosen.

In this framework, we built solutions to some well-known numerical **NP**-complete decision problems: the subset sum and the knapsack problems. An implementation of these cellular solutions, using our Prolog simulator, is also discussed.

The presented software can be used to deal with other solutions to **NP**-complete problems that use active membranes. It suffices to provide the auxiliary generator tool included in our simulator with the appropriate skeleton of the solution (or, alternatively, to introduce *ex profeso* the system to be simulated).

Although it is useful on many counts, the presented simulator is not efficient from the computational point of view. First, it is designed to run on a single sequential processor. Second, the instances of the problem are codified in 1-ary form, via multisets. It seems rather natural that one of the future improvements of the simulator will be adapting the tool to parallel Prolog.

Nevertheless, the current simulator has proven to be very useful for debugging the process of formal verification of cellular designs. And since it supplies a comprehensive description of P system evolutions, we believe that it is quite suitable for educational purposes. In this direction, we are currently working on developing a user-friendly variant with a graphical interface.

ACKNOWLEDGMENTS

Support for this research through the project TIC2002-04220-C03-01 of the Ministerio de Ciencia y Tecnología of Spain, cofinanced by FEDER funds, is gratefully acknowledged.

REFERENCES

Adleman, L. M. (1994). Molecular computations of solutions to combinatorial problems. *Science*, 226, 1021-1024.

Bratko, I. (2001). *PROLOG programming for artificial intelligence*. Boston: Addison-Wesley.

Cordón Franco, A., Gutiérrez Naranjo, M. A., Pérez Jiménez, M. J., Riscos Núñez, A., & Sancho Caparrini, F. (2004). Implementing in Prolog an effective cellular solution to the knapsack problem. In Martín-Vide, C., Mauri, G., Păun, Gh., Rozenberg, G., & Salomaa, A. (Eds.), *Membrane computing. International Workshop, WMC 2003*, Tarragona, Spain, revised papers. *Lecture Notes in Computer Science*, 2933, 140-152.

Head, T. J. (1987). Formal language theory and DNA: An analysis of the generative capacity of specific recombinant behaviors. *Bulletin of Mathematical Biology*, 49, 737-759.

Holland, J. H. (1975). *Adaptation in natural and artificial systems*. Ann Arbor, MI: University of Michigan Press.

McCulloch, W. S. & Pitts, W. (1943). A logical calculus of the ideas immanent in nervous activity. *Bulletin of Mathematical Biophysics*, 5, 115-133.

Păun, Gh. (1998). Computing with membranes. Turku Center for Computer Science *TUCS Report*, 208. Also in (2000) *Journal of Computer and System Sciences*, *61* (1), 108-143.

Păun, Gh. (2001). P Systems with active membranes: Attacking NP-complete problems. *Journal of Automata, Languages and Combinatorics*, *6* (1), 75-90.

Păun, Gh. (2002). *Membrane computing: An introduction*. Berlin: Springer.

Pérez Jiménez, M. J., Romero Jiménez, A., & Sancho Caparrini, F. (2004). The **P** versus **NP** problem through cellular computing with membranes. *Aspects of Molecular Computing. Lecture Notes in Computer Science*, 2950, 338-352.

Zandron, C. (2001). *A model for molecular computing: Membrane systems*. Ph.D. Thesis, Università degli Studi di Milano.

APPENDIX

In what follows, we show some of the rules generated by the simulator for the instance of the problem considered. Note that the number after ** is the ordinal of the corresponding rule.

```
% Set (R1)
p1 rule e0 inside_of e ec 0 divides_into q inside_of e ec -1
   and e0 inside_of e ec 1 ** 1.
      ...
```

p1 rule e4 inside_of e ec 0 divides_into q inside_of e ec -1
and e4 inside_of e ec 1 ** 5.

p1 rule e0 inside_of e ec 1 divides_into e1 inside_of e ec 0
and e1 inside_of e ec 1 ** 6.

...

p1 rule e3 inside_of e ec 1 divides_into e4 inside_of e ec 0
and e4 inside_of e ec 1 ** 9.

% Set (R2)
p1 rule x0 evolves_to [a0_] in e ec 0 ** 10.

p1 rule y0 evolves_to [b0_] in e ec 0 ** 11.

p1 rule x0 evolves_to [] in e ec 1 ** 12.

p1 rule y0 evolves_to [] in e ec 1 ** 13.

p1 rule x1 evolves_to [x0] in e ec 1 ** 14.

...

p1 rule x4 evolves_to [x3] in e ec 1 ** 17.

p1 rule y1 evolves_to [y0] in e ec 1 ** 18.

...

p1 rule y4 evolves_to [y3] in e ec 1 ** 21.

% Set (R3)
p1 rule q evolves_to [q_, q0] in e ec -1 ** 22.

p1 rule b0_ evolves_to [b0g] in e ec -1 ** 23.

p1 rule a0_ evolves_to [a0] in e ec -1 ** 24.

p1 rule b_ evolves_to [bg] in e ec -1 ** 25.

p1 rule a_ evolves_to [a] in e ec -1 ** 26.

% Set (R4)
p1 rule a0 inside_of e ec -1 sends_out # of e ec 0 ** 27.

p1 rule a inside_of e ec 0 sends_out # of e ec -1 ** 28.

% Set (R5)
p1 rule q0 evolves_to [q1] in e ec -1 ** 29.

p1 rule q2 evolves_to [q3] in e ec -1 ** 30.

   ...
p1 rule q1 evolves_to [q2] in e ec 0 ** 33.

p1 rule q3 evolves_to [q4] in e ec 0 ** 34.

p1 rule q5 evolves_to [q6] in e ec 0 ** 35.

% Set (R6)
p1 rule q1 inside_of e ec -1 sends_out # of e ec 1 ** 36.

p1 rule q3 inside_of e ec -1 sends_out # of e ec 1 ** 37.

p1 rule q5 inside_of e ec -1 sends_out # of e ec 1 ** 38.

p1 rule q7 inside_of e ec -1 sends_out # of e ec 1 ** 39.

% Set (R7)
p1 rule q_ evolves_to [q_0] in e ec 1 ** 40.

p1 rule b0g evolves_to [b0] in e ec 1 ** 41.

p1 rule bg evolves_to [b] in e ec 1 ** 42.

p1 rule a evolves_to [] in e ec 1 ** 43.

% Set (R8)
p1 rule b0 inside_of e ec 1 sends_out # of e ec 0 ** 44.

p1 rule b inside_of e ec 0 sends_out # of e ec 1 ** 45.

% Set (R9)
p1 rule q_0 evolves_to [q_1] in e ec 1 ** 46.

p1 rule q_2 evolves_to [q_3] in e ec 1 ** 47.

p1 rule q_4 evolves_to [q_5] in e ec 1 ** 48.

p1 rule q_6 evolves_to [q_7] in e ec 1 ** 49.

p1 rule q_8 evolves_to [q_9] in e ec 1 ** 50.

p1 rule q_1 evolves_to [q_2] in e ec 0 ** 51.

p1 rule q_3 evolves_to [q_4] in e ec 0 ** 52.

p1 rule q_5 evolves_to [q_6] in e ec 0 ** 53.

p1 rule q_7 evolves_to [q_8] in e ec 0 ** 54.

% Set (R10)
p1 rule q_9 inside_of e ec 1 sends_out yes of e ec 0 ** 55.

p1 rule q_9 inside_of e ec 0 sends_out yes of e ec 0 ** 56.

% Set (R11)
p1 rule z0 evolves_to [z1] in s ec 0 ** 57.

p1 rule z1 evolves_to [z2] in s ec 0 ** 58.
          ...
p1 rule z26 evolves_to [z27] in s ec 0 ** 83.

p1 rule z27 evolves_to [z28] in s ec 0 ** 84.

p1 rule z28 evolves_to [d1, d2] in s ec 0 ** 85.

% Set (R12)
p1 rule d1 inside_of s ec 0 sends_out d1 of s ec 1 ** 86.

p1 rule d2 evolves_to [no] in s ec 1 ** 87.

p1 rule yes inside_of s ec 1 sends_out yes of s ec 0 ** 88.

p1 rule no inside_of s ec 1 sends_out no of s ec 0 ** 89.

Chapter VI

Modeling Developmental Processes in MGS

Jean-Louis Giavitto,
CNRS – University of Évry Val d'Essonne – Genopole, France

Olivier Michel,
University of Évry Val d'Essonne – Genopole, France

ABSTRACT

Biology has long inspired unconventional models of computation to computer scientists. This chapter focuses on a model inspired by biological development both at the molecular and cellular levels. Such biological processes are particularly interesting for computer science because the dynamic organization emerges from many decentralized and local interactions that occur concurrently at several time and space scales. Thus, they provide a source of inspiration to solve various problems related to mobility, distributed systems, open systems, etc. The fundamental mechanisms of biological development are now understood as changes within a complex dynamical system. This chapter advocates that these fundamental mechanisms, although mainly developed in a continuous framework, can be rephrased in a discrete setting relying on the notion of rewriting in a topological setting. The discrete formulation is as formal as the continuous one, enables the simulation, and opens a way to the

*systematic study of the behavioral properties of the biological systems.
Directly inspired from these developmental processes, the chapter presents
an experimental programming language called MGS. MGS is dedicated to
the modeling and simulation of dynamical systems with dynamical
structures. The chapter illustrates the basic notions of MGS through
several algorithmic examples and by sketching various biological models.*

INTRODUCTION

The membrane paradigm, DNA computing, molecular computing, and aqueous computing are examples of unconventional models of computation inspired by molecular biology. In this chapter, we focus on a model inspired by biological development at both molecular and cellular levels. We are interested not only in the interactions between the molecules, but also by the assembling and the structural organization that is dynamically created.

Such biological processes are particularly interesting for a computer scientist because the dynamic organization of the involved entities emerges from many decentralized and local interactions that occur concurrently at several time and space scales. The development of biological organisms has for a long time inspired computer science (see, for instance, the notion of cellular automata noted in von Neumann, 1966). More recently, the emerging domains of *amorphous computing* (*www.swiss.ai.mit.edu/projects/amorphous*), *self-healing systems,* or *autonomic computing* (Kephart & Chess, 2003; Parashar & Hariri, 2003) are also directly inspired by developmental processes found at both molecular and cellular levels. Inspired by the description of various developmental processes as changes in a dynamical system, we propose a new computational paradigm that extends the idea of rewriting systems to a broader class of data structures. To investigate this model, we are developing an experimental programming language called MGS.

However, the fertilization of computer science by biological notions (Paton, 1994) is not a one-way process, and biology has imported many concepts developed within computer science such as the notion of programs, memory, information, control, and many others (Stengers, 1988; Keller, 1995). Obviously, new programming paradigms inspired by basic developmental mechanisms will be also an ideal framework to support and help the biologist in analysing and understanding these kinds of biological processes. We illustrate this cross-fertilization by using MGS to simulate various processes of pattern formation.

Outline of the Chapter

In the next section, we will sketch the requirements needed to model developmental processes at both molecular and cellular levels. Our approach to modeling developmental processes is based on the notion of rewriting. This is not really a novelty; although most of the models in this field rely on partial differential equations, string rewriting through the notion of L systems (Lindenmayer, 1968) is now routinely used to model the development of plants, and multiset rewriting constitutes the theoretical foundation of artificial chemistries (*http://ls11-www.cs.uni-dortmund.de/achem*). The section "A Very Short Introduction to Rewriting Systems" presents some fundamental notions of rewriting with an emphasis on the use of rewriting rules to specify the evolution function of a dynamical system.

The usual notion of term rewriting is too restrictive to be easily used to model biological entities. This drawback motivates the development of MGS, an experimental programming language devoted to the simulation of dynamical systems with a dynamical structure. The section "A Quick Presentation of MGS" sketches the basic principles of MGS and outlines how several rewriting mechanisms are unified in the same framework using the notions of *topological collection* and *transformation*. The presentation is restricted to the notions required to understand the examples given in the next two sections.

The next section, entitled "Paradigms of Pattern Formation", focuses on biological system modeling and details some examples of fundamental processes involved in pattern formation:

- Diffusion and beyond (Fick's law and molecular diffusion).
- Boundary growth (the growth of a snowflake and diffusion limited aggregation).
- Growth of a tumor (illustrating the coupling between mechanical forces and cell division).

The examples provided show the ability of MGS to express, in a simple and straightforward way, various patterns of development. They also validate our approach for the modeling and the simulation of dynamical systems with a dynamical structure. A comparison of our approach to related works and some other models of computation concludes the chapter.

BACKGROUND: DEVELOPMENT AND DYNAMICAL SYSTEMS

A Dynamical System with a Dynamical Structure

In the field of developmental biology, one current theoretical framework views the developmental process as changes within a dynamical system (DS). This point of view can be traced back at least to D'Arcy Thompson, Turing, von Bertalanffy, and Waddington, and contrasts with pure genetically programmed and pre-existing plans that await revelation during the developmental process. In the past two decades or so, the concepts and models of nonlinear dynamical systems have been coupled with models of genetic regulations to overcome the genetic/epigenetic debate on the nature of the ontogenetic processes. These models can be seen in the pioneering works of researchers such as Harper et al. (1986), Kaufman (1995), Maynard-Smith (1999), Varela (1979), Wolpert et al. (2002), and Meinhardt (1982).

A developmental process viewed as a dynamical system often presents the distinctive feature of having a dynamic phase space (the phase space is the set of all possible states of the system). Consider a "classical" dynamical system, like a falling stone. This system is adequately described by a position and a velocity. Even if the position and the velocity of the stone are constantly changing in time, the state of this system is always given as a vector in $\mathbf{R}^3 \times \mathbf{R}^3$. We see that the phase space is given *a priori*, and we say that the DS has a static structure.

In contrast, consider the development of an embryo. Initially, the state of the system is described solely by the chemical state c_0 of the egg (no matter how complex this chemical state). After several divisions, the state of the embryo must describe the chemical states c_i of the old and new cells *and also* their spatial arrangement. The number of cells and their organization at a given time cannot be given *a priori*. Moreover, there is a kind of "circular causality" in the evolution of the chemical states and the evolution of the spatial arrangement of the cells: molecules interfere with the placement of the cells (because some molecules make the membranes sticky) and the position of the cells interferes with the diffusion of the molecules (the cells are the medium of the diffusion and create compartments that change the diffusion). Consequently, the exact phase space of the system cannot be fixed before the evolution and development models must state the evolution of the spatial structure jointly with the evolution of the cells states. We say that this kind of DS exhibits a dynamical structure.

In the rest of this chapter we use the abbreviation $(DS)^2$ to mean "dynamical system with a dynamical structure" (Giavitto & Michel, 2003).

The importance of the dynamical structure of a biological system perceived as a dynamical system has often been recognized under several names. *Hypercycle* in autocatalytic networks (Eigen & Schuster, 1979), *autopoeisis* (Varela, 1979), *variable structure systems* in classical control systems (Itkis, 1976; Hung et al., 1993), *developmental grammar* (Mjolsness et al., 1991), and *organization* (Fontana & Buss, 1994) are notions that have been developed to catch and formalize this idea of a changeable structure of a system.

As a matter of fact, the specification of the evolution of a $(DS)^2$ can be very difficult to achieve, and new programming concepts must be developed to help modeling and simulation. Indeed, standard approaches in the formalization of DS do not allow the specification of the phase space or its topology as an observable of the system[1]. This observation has motivated the development of the MGS project.

Towards a Discrete Conceptual Framework

As in many other complex systems in nature, $(DS)^2$ can exhibit coherent behaviors — the parts are coordinated locally and without a centralized control to produce an emergent, organized, stable, and robust pattern. Here, by pattern, we mean the forms or the shapes produced by the development processes in space or in time (Stevens, 1974). For example, the periodic occurrence of a predefined event, such as a chemical concentration repeatedly reaching a given level before decreasing and increasing again, is also an example of a pattern, but in time rather than in space.

What conceptual framework could be used to specify and analyze these developmental design patterns? A quick look at the works in this area shows that although discrete formalizations are present, the mainstream of the contributions are developed in the framework of continuous models. For example, the famous reaction-diffusion model introduced by A. Turing (1952) is described by a set of partial differential equations (PDEs). However, we advocate that discrete models can be used advantageously to model biological development for several significant reasons:

- In the discrete approach, the composition of the design patterns can be studied from an *algebraic* point of view and, more precisely, from *combinatorial* and *generative* points of view to analyze the space of possible patterns. This last approach profits from all of the results and tools developed in the field of formal language theory and is able to handle the changes of phase space during the evolution.

- The discrete approach is definitely more abstract — it focuses on the fundamental functional and algorithmic properties of pattern formation, without the burden of the "physico-chemical implementation".
- The discrete approach focuses on high-level properties and is well adapted to our knowledge of the biological data. For example, actual DNA chip measurements usually distinguish only between three and eight relevant activation levels for a gene, metabolic fluxes cannot generally be measured *in vivo* for only one cell, etc. In addition, it is more fruitful to have a functional description (e.g., "the cell is in this phase") than a precise quantitative description ("the concentration of this chemical has reached this level"). In fact, the functional description is at last what we are looking for, and the link between the two kinds of descriptions is highly dependent on the experiment, the organism, the environment, etc.
- Development is often described as a succession of discrete morphogenetic events that represent a discontinuity (a cell divides and we have to handle the change of one cell description to two cell descriptions).
- Finally, the continuous framework raises crippling difficulties:

    - Since the equations involved often have nonlinear terms, they can be solved only by numerical methods.
    - The question of the robustness of the solutions of PDEs is completely open, mainly due to the lack of qualitative understanding of such equations.
    - The continuous formalism makes it difficult to express the discrete nature of the biological entities.

From the modeling perspective, the discrete approach is too abstract, leaving unspecified the actual embodiment of the pattern formation. However, despite the increasing amount of biological data, we are far from obtaining the quantitative data needed by realistic continuous models.

So, at this point we are looking for a formal discrete framework able to support the specification of developmental design patterns. At last, the developmental process consists of transformations of the system's parts, if by transformation we understand the appearance and the disappearance of matter and the changes in quality (size, differentiation) or in position of this matter. Then, the global changes of the whole system must be specified as several local competing transformations occurring in an organized set of simpler entities. This idea of replacing some parts of a structure by other parts specified by local rules recalls the notion of *rewriting*.

A VERY SHORT INTRODUCTION TO REWRITING SYSTEMS

Based on the notion of rewriting systems (Dershowitz & Jouannaud, 1990; Dershowitz, 1993), the emphasis of this section is on the concepts that are relevant in the context of applying rewriting systems to the simulation of dynamical systems (Giavitto 2003; Giavitto et al., 2004).

A Computational Device

Computer science has appropriated and developed the notion of rewriting systems (RS) to build and relate different processes. The mechanics of rewriting systems are familiar to anyone who has done high school mathematics: a term can be simplified by repeatedly replacing parts of the term (i.e., subterms) with other equivalent subterms. Terms represent expressions by trees — each internal node is labeled by an operation and the leaves are labeled by constants. A subterm replacement is specified as a rule $\alpha \rightarrow \beta$ where the left-hand-side α specifies a subterm, and the right-hand-side β specifies its replacement. A variable x in α is a kind of wildcard that specifies an entire and unknown subterm; a variable x in β denotes the subterm matched by x in α.

An example involving arithmetic expressions is illustrated in Figure 1. An arithmetic expression is adequately represented by a tree in which the internal nodes are labeled by arithmetic operators and the leaves by integer numbers. An expression with variables is a filter that matches a subtree, the variables catching an entire subtree. A rewrite rule $\alpha \rightarrow \beta$ specifes that a subtree of a given form α must be replaced by a tree β. Several occurrences of a subtree may need to be replaced. In the top-left tree there are two occurrences of a subtree matched by $0+x$. The *rewriting strategy* indicates which occurrence must be replaced. The result of the three possible replacements (replacing one occurrence, the other, or the two) are figured at the right side of the Figure 1.

Let R be a given set of rules, and e a term: we write $e \rightarrow_R e'$ to denote that e can be rewritten in e' using one application of a rule of R. We omit the index R when there is no ambiguity regarding the set of rules used, and we write $\rightarrow^*$ to denote the reflexive-transitive closure of the relation $\rightarrow$. A sequence $e_1 \rightarrow$

Figure 1. Rewriting an arithmetic expression

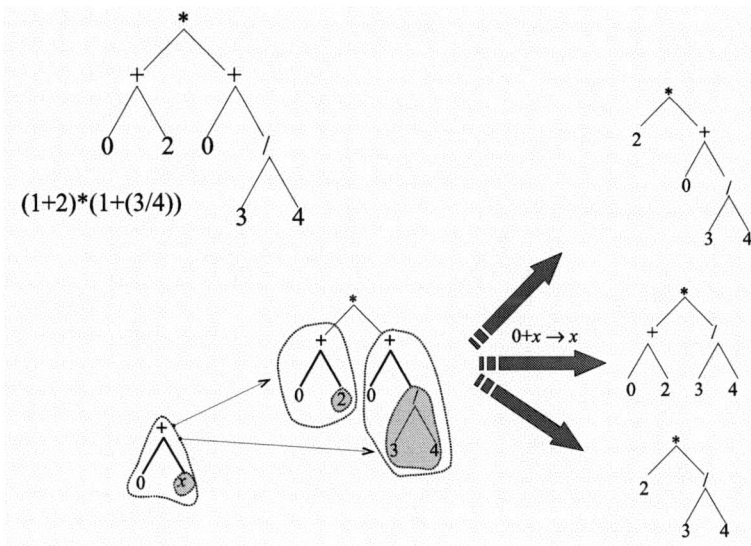

$e_2 \rightarrow ... \rightarrow e_n$ is called a derivation. The final result of an unextendible sequence of rule applications (i.e., e_n cannot be further rewritten) is called a *normal form*.

RS can be viewed as a kind of directed equations used to compute by repeatedly replacing subterms of a given formula with equal terms until the simplest possible form is obtained. This device has been primarily used as a decision procedure in equational theory: if we can prove that the normal form computed by a RS is unique, then two terms can be proven to be equivalent (modulo the equations corresponding to the rules of the RS) if they share the same normal form. As a formalism, RS has the full power of Turing machines and may be thought of as nondeterministic Markov algorithms over terms, rather than strings.

Rewriting and Simulation

The previous description strongly suggests that a rule $\alpha \rightarrow \beta$ can be used to represent the local evolution of a subsystem α into a new state β. That is, RS can be used for the modeling of a DS, provided that:

- The state of the DS is represented by a term and a subterm represents the state of a subsystem.
- The evolution function of the DS can be specified as the rules of the RS.

For instance, in the context of the development of the embryo, a rule $c \oplus i \rightarrow c' \oplus i'$ can be used to specify that a cell in state c that receives a signal i evolves in state c' and sends the signal i'. A rule such as $c \rightarrow c' \oplus c''$ defines a cellular division and $c \rightarrow \emptyset$ (c gives nothing) represents apoptosis. The idea is that the left-hand-side of a rewriting rule selects an entity in the biological system and the messages that are addressed to it, and the right-hand-side describes the new state of the entity and the eventual messages that are emitted. The operator $\oplus$ that appears in the rule denotes the composition of the entities together with the messages into an entire global system. The ability to equally express the change of state and the appearance or disappearance of entities makes RS suitable to model $(DS)^2$.

The Management of Time

The notion of time that underlies the use of RS for the modeling of DS is clearly based on a discrete atomic event model: time is passing when some event occurs somewhere in the system, a rule application corresponds to the occurrence of such an event, and duration is not handled (but can be emulated using a start and an end event).

A *rewriting strategy* is an algorithm for choosing in a term e the occurrence of the subterm that must be rewritten. Several algorithms can be considered that allow some control over the management of time. For example, intuitively, the laws of nature are encoded into the rules of an RS and these laws must apply everywhere. This leads to consideration of the so-called "parallel application strategy". Instead of considering just one occurrence of a rule application during a rewriting step, one may chose to rewrite in the same step a maximal nonintersecting set S of matching subterms in e: two subterms in S do not have a subterm in common, each element in S matches the left-hand-side of a rule, and S cannot be further extended by a subterm in e without losing the two previous properties. On the other hand, one may suppose that each event occurs at a different time. This asynchronous dynamic fits well with the standard application of only one rule at each rewriting step.

The Management of Space

A rule of form $c \oplus i \rightarrow c' \oplus i'$ assumes that signal i produced by some cell reaches its target c, and that signal i' produced by the change of c to c' reaches its target located somewhere in the system. Thus, the operation $\oplus$ used to compose the state of the subsystems and the interaction messages must be able to express the spatial localization and the functional organization of the system.

For instance, suppose that our cells are bacteria in a test tube. The signals produced by one cell are then released in a solution. By thermal agitation, the signals potentially reach any other cell in the test tube. This can easily be achieved by using a (formal) operator $\oplus$ that is *associative* and *commutative*. From this point of view, the term $c_1 \oplus c_2 \oplus c_3$ represents a test tube with three cells. Because we assume that $\oplus$ is associative and commutative, this term can be rewritten in the equivalent term $c_2 \oplus c_1 \oplus c_3$ or $c_3 \oplus c_1 \oplus c_2$. Applying the rule $c_1 \rightarrow c_1 \oplus i$ to this term gives $c_1 \oplus i \oplus c_2 \oplus c_3$. By associativity and commutativity, signal i can be moved to the neighbor of c_3 or any other cell.

Terms built with associative and commutative operators achieve a multiset organization of the objects. A multiset is a set in which an element is allowed to occur multiple times. A multiset builds up a "chemical soup" of elements that are not bound in a tree but can move around and interact with one another.

By imposing only associativity, the term structure reduces to a sequence of elements. So, by giving some properties to the operations in the term, we can represent several kinds of organization. This fact has long been recognized, and multiset rewriting and string rewriting have been successfully used in the field of biological modeling. The rest of this section gives some references of these approaches.

However, multisets, sequences, and trees of elements are far from being sufficient to characterize the sophisticated organization needed to represent the variety of biological structures from molecules to societies, through compartments, cells, tissues, organs, and individuals. This severe shortcoming motivates (among others reasons) the extension of the notions of rewriting to more general structures. In the next section we present such an extension based on topological notions.

Multiset Rewriting

Multiset rewriting is the core of the Gamma language. Gamma (Banâtre & Le Metayer, 1986; Banâtre et al., 1987) is based on the chemical reaction metaphor; the data are considered as a multiset M of molecules and the computation is a succession of chemical reactions according to a particular rule. A rule (R, A), where R is a predicate and A a function, indicates which kind of molecules can react together (a subset m of M that satisfies predicates R) and the product of the reaction (the result obtained by applying function A to m). Several reactions may happen in the same time. No assumption is made on the order in which the reactions occur. The only constraint is that if the reaction condition R holds for at least one subset of elements, at least one reaction occurs (computation stops once the reaction condition does not hold for any subset of the multiset). The CHemical Abstract Machine (CHAM) extends these ideas with a focus on the semantics of nondeterministic processes (Berry & Boudol, 1992).

The application of this abstract chemistry-based approach is now recognized as an emerging field called *artificial chemistries* (Dittrich et al., 2001; *http://ls11-www.cs.uni-dortmund.de/achem*). Various motivations presides the growing body of research done in this field, ranging from the study of the automated generation of combustion reactions (Bournez et al., 2003) to the study of complex dynamical systems and self-organization in biological evolution (Fontana & Buss, 1994).

An important application is the modeling and the study of the behavior of signaling pathways. Fisher et al. (2000) proposed the use of rewriting systems to model cascades of protein interactions in signaling pathways. Later work (Eker et al., 2002; Lincoln, 2003) has produced some very sophisticated models of these pathways; however, the earlier work draws attention to the subtle role that so-called scaffold proteins play in facilitating cascades and preventing cross-talk between pathways. The RS approach provides extremely efficient representations of the information processing nature of signaling pathways. Despite the obvious use of RS for simulating various behaviours, the symbolic tool sets produced (e.g., model checking) can then be applied to generate and check novel hypotheses and to develop an algebra and a logic of signaling pathways.

String Rewriting

If the operator involved in the term to be rewritten is only associative (and not commutative), then the term simply corresponds to a sequence of elements. Such sequences are also called *strings* (especially if the universe of possible elements is finite). String rewriting is a formal framework heavily investigated, for example, in formal language theory.

String rewriting has shown its usefulness and maturity in plant development modeling. Introduced by Lindenmayer (1968), the L system formalism can be roughly described as string rewriting rules applied in parallel to strings representing a linear or a branching structure. The original L system formalism has been extended in many ways, and comprehensive reviews have been produced (Prusinkiewicz & Lindenmayer, 1990; Prusinkiewicz 1998, 1999). It is worth giving a flavor of this formalism through a very simple example.

The two rules, $a \rightarrow ab$ and $b \rightarrow a$ can be viewed as a model of the development of a filamentous organism using the symbols a and b. The first rule states that a cell of length a divides into two adjacent cells of length a and b. The other rule states that a cell b changes its length to a (these rules are related to the development of the bacterium *Anabaena* where we have ignored the polarity that would determine if a cell of length a divides into ab or ba). The sequence of the cells is simply denoted by the juxtaposition of their symbols. Starting from a unique cell of length a, we obtain successively:

$$a \rightarrow ab \rightarrow aba \rightarrow abaab \rightarrow abaababa \rightarrow abaababaabaab \rightarrow \ldots$$

by applying the two rules in parallel. Each word represents a state of the development of the filament.

A good example of an L system that takes into account the cellular interaction in the development is the modeling of growth and heterocyst differentiation in *Anabaena* (Wilcox et al., 1973; Hammel & Prusinkiewicz, 1996). This example is remarkable for at least two reasons: it shows the ability of this kind of discrete model to accommodate or to easily emulate features usually handled in continuous formalisms (e.g., the modeling of the diffusion) and also because it tackles a fundamental biological mechanism — a morphogenesis driven by a reaction-diffusion process taking place in a growing medium.

By combining and structuring multiset and string rewriting, we can extend the applicability of these formalisms. Applications of such extensions at the genetic level include DNA computing (Adleman, 1994) and splicing systems, a language-theoretic model of DNA recombination that allows the study of the generative power of general recombination and of sets of enzymatic activities (Head, 1987, 1992).

A QUICK PRESENTATION OF MGS

We present the fundamental notions that underlie the MGS programming language. MGS stands for "encore un Modèle Général de Simulation" (or, "yet another general model for simulation"). The notion of *topological collection* developed in MGS enables the unification of various forms of rewriting and its extension to more general data structures than trees.

Topological Collection and Their Transformations

MGS is aimed at the representation and manipulation of local transformations of entities structured by *abstract topologies* (Giavitto & Michel, 2001,

2002). A set of entities organized by an abstract topology is called a *topological collection*. "Topological" here means that each collection type defines a neighborhood relation specifying both the notion of *locality* and the notion of *subcollection*. A subcollection *B* of a collection *A* is a subset of elements of *A* defined by some *path* and inheriting its organization from *A*. A path is a sequence of adjacent elements, the adjacency relation being specified by the neighborhood relationship between the elements of the data structure.

Abstractly, a topological collection can be formalized as a partial function *C* that associates a value to the points of a discrete topological space. The points of this space are called the *positions* of the collection. The values associated with these positions, or the image of the partial function *C,* are the elements of the collection. The topological structure associated with the set of positions gives the neighborhood relationship between the collection elements. For example, a 2-D array filled with integer elements can be seen as a partial function from $\mathbf{Z}^2$ to $\mathbf{N}$ with a finite definition domain. Here, the topology of $\mathbf{Z}^2$ is inherited from the module structure of $\mathbf{Z}^2$ (Munkres, 1993) and a position (x, y) is the neighbor of a position (x',y') only if $x'=x\pm1$ and $y'=y\pm1$. This representation of an array is only an abstract view — an array is not really implemented as a function within the computer but as a set of values indexed by positions taken in $\mathbf{Z}^2$.

The *global* transformation of a topological collection *C* consists of the parallel application of a set of *local* transformations. A local transformation is specified by a rewriting rule *r* that specifies the change of a subcollection. The application of a rewrite rule $r = \alpha \to f(\alpha)$ to a collection *A:*

- Selects a subcollection *B* of *A* whose elements match the path pattern α.
- Computes a new collection *C* as a function *f* of *B* and its neighbors.
- Specifies the insertion of *C* in place of *B* into *A*.

In other words, MGS extends the idea of the term by the idea of topological collection and generalizes the notion of the rewriting rule to the notion of transformation. For the sake of the expressivity, MGS embeds the idea of topological collections and their transformations into the framework of a simple dynamically typed functional language. Collections are just new kinds of values, and transformations are functions acting on collections and defined by a specific syntax using rules. Functions and transformations are first-class values and can be passed as arguments or returned as the result of an application. MGS is an applicative programming language: operators acting on values combine values to give new values, they do not act by side-effect. In our context, *dynamically typed* means that there is no static type checking and that type errors are detected at run-time during evaluation. Although dynamically typed, the set of values has a rich type structure used in the definition of pattern-matching, rules, and transformations.

Collection Types

There are several predefined collection types in MGS, and also several means of constructing new collection types. The collection types can range in MGS from totally unstructured with sets and multisets to more structured with sequences and GBFs (GBFs generalize the notion of regular array and are presented in the section entitled "Group-Based Data Field"). Other topologies are currently under development and include general graphs and abstract simplicial complexes. Abstract simplicial complexes generalize the notion of graphs and enable the representation of arbitrary topology (Munkres, 1993). However, in this chapter we are mainly concerned with three families of collection types: *monoidal collections*, *GBFs*, and *graphs*.

For any collection type T, the corresponding empty collection is written $():T$. The name of a type is also a predicate used to test if a value v has this type: $T(v)$ holds only if v is of type T. Each collection type can be subtyped:

    collection U = T

introduces a new collection type U, which is a subtype of T. These two types share the same topology but a value of type U can be distinguished from a value of type T by the predicate U. Elements in a collection T can be of any type, including collections, and thus achieving complex objects in the sense of Buneman et al. (1995).

Monoidal Collections

Sets, multisets, and sequences are members of the monoidal collection family. In fact, a sequence (a multiset, a set) of values taken from V can be seen as an element of the free monoid V^* (the commutative monoid, the idempotent, and commutative monoid, respectively). The join operation in V^* is written by a comma operator "," and induces the neighborhood of each element: let E be a monoidal collection, then the element y in E is the neighbor of the element x iff $E=u,x,y,v$ for some u and v. This definition induces the following topology:

- For sets (type **set**), each element in the set is a neighbor of any other element (because of the commutativity, the elements in a term describing a set can be reordered following any order and they are all distinct because of the idempotent property).
- For multisets (type **bag**), each element is also a neighbor of any other one (however, the elements are not required to be distinct as in a set).
- For sequences (type **seq**), the topology is the expected one — an element not at the right end has a neighbor at its right.

The comma operator is overloaded in MGS and can be used to build any monoidal collection (the type of the arguments disambiguates the collection built). So, the expression 1,1+1, 2+1,():**set** builds the set with the three elements 1, 2 and 3, while expression 1,1+1,2+1,():**seq** makes a sequence *s* with the same three elements. The comma operator is overloaded such that if *x* and *y* are not monoidal collections, then x,y builds a sequence of two elements. So, the expression 1,1+1,2+1 evaluates to the sequence *s* as well.

Group-Based Data Field

Group-based data fields (GBF for short) are used to define organizations with *uniform* neighborhood. A GBF is an extension of the notion of array, where the elements are indexed by the elements of an abelian group, called the *shape* of the GBF (Giavitto et al., 1995; Michel 1996; Giavitto 2001). For example:

gbf Grid2 = < north, east >

defines a GBF collection type called **Grid2**, corresponding to the von Neumann neighborhood in a classical array (a cell above, below, left, or right, but not diagonal). The two names **north** and **east** refer to the directions that can be followed to reach the neighbors of an element. These directions are the *generators* of the underlying group structure. The inverse of the generators can also be followed to reach a neighbor. The right-hand side (r.h.s.) of the GBF definition gives a finite presentation of the group structure. The list of the generators can be completed by giving equations that constraint the displacements in the shape:

gbf Hex = <east, north, northeast; east+north = northeast>

defines a hexagonal lattice that tiles the plane, as shown in Figure 2.

In this diagram, an hexagonal cell represents a group element and neighbors' elements share a common edge. Each cell has six neighbors (following the three generators and their inverses). The equation east+north = northeast specifies that a move following northeast is the same as a move following the east direction followed by a move following the north direction. This representation can be easily generalized to visualize the topology of any GBF of type T by a graph. The result is the *Cayley graph* of the presentation of T: each vertex in the Cayley graph is an element of the group, and vertices x and y are linked if there is a generator g in the presentation such that $x+g=y$. This representation enables a dictionary between graph theoretic notions and group concepts.

A word (a formal sum of the group generators) is a path in the Cayley graph. Path composition corresponds to group addition, and equation $P+v=Q$, where P and Q are given, always has a solution: two cells in the graph are always connected. Each cell c can be named by the words that represent a path starting from the cell 0 and ending in c. All of these words represent the same group element. A closed path (a cycle) is a word equal to 0 (the identity of the group operation). There are two kinds of cycles in the graph. The first ones are present in all groups and correspond to group laws (intuitively, a backtracking path such as $b+a \ominus a \ominus b$ where a and b are generators). The other closed paths are specific to the group equations. An equation $v=w$ can be rewritten $v \ominus w = 0$ and, thus, corresponds to a closed path. In the diagram, the closed triangular path on the top left corresponds to the equation of the GBF, and the closed path on the top right corresponds to the commutation of the generators east and north. See Figure 2 for an illustration.

Figure 2. Shapes of a GBF <north, east, northeast; east+north = northeast>

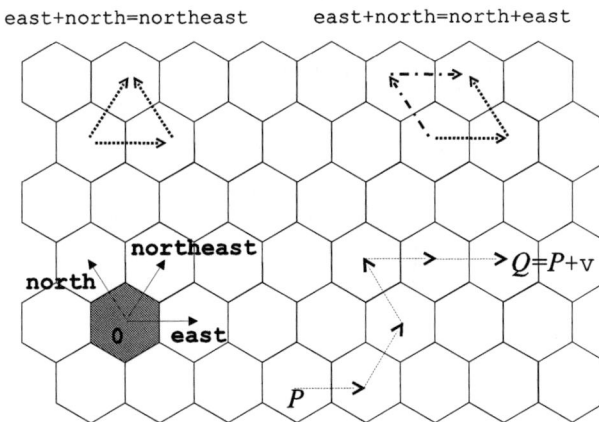

A GBF value of type T is a *partial function* that associates a value to some group elements (the group elements are the positions of the collection and the empty GBF is the everywhere undefined function). A GBF value is simply a labeling of a finite set of positions by some values. The positions that have no values are said to be *undefined*.

Matching a Path

Path patterns are used in the left-hand side (l.h.s.) of a rule to match a subcollection to be substituted. We give only a fragment of the grammar of the patterns:

Pat ::= *x* | <undef> | *p,p'* | *p* |g> *p* | *p*+ | *p*/*exp* | *p* as *x*

where *p, p'* are patterns, g is a GBF generator, *x* ranges over the pattern variables, and *exp* is an expression evaluating to a Boolean value. Informally, a path pattern can be flattened into a sequence of basic filters and repetition specifying a sequence of positions with their associated values. The order of the matched elements can be forgotten to see the result of the matching as a subcollection.

A pattern variable *x* matches exactly one element somewhere in the collection that has a well-defined value. The identifier *x* can be used in the rest of the rule to denote the value of the matched element. More generally, the naming of the values of a subpath is achieved using the construction **as**. The constant <undef> is used to match an element with an undefined value (i.e., a position in a topological collection with no value). The pattern *p,p'* stands for a path beginning like *p* and ending like *p'* (i.e., the last element in path *p* must be a neighbor of the first element in path *p'*). For example, the pattern a,b matches two connected elements referred to hereafter as a and b (i.e., b must be a neighbor of a).

The neighborhood relationship depends on the collection kind and is decomposed in several subrelations in the case of a GBF. The comma operator is then refined in the construction *p* |g> *p'*: the first element of *p'* is the g-neighbor of the last element in path *p*. The pattern *p*+ matches a repetition *p, ..., p* of path *p*. Finally, *p*/*exp* matches the path *p* only if *exp* holds.

Here is a more contrived example:

(e/seq(e))+ as S / size(S)<5

selects a subcollection S of less than five elements, each element e of S being a sequence. If this pattern is used against a set, S is a subset; if this pattern is used against a sequence, S is a subsequence (that is, an interval of contiguous elements), etc.

Path Substitution and Transformations

There are several features to control the application of a rule. Rules may have priorities or probabilities of application, they may be guarded and depend on the value of local variables, they can "consume" their arguments, etc. We present only the basic application strategy; see Giavitto and Michel (2001) for more details.

Substitutions of Subcollections

A rule $\alpha \rightarrow \beta$ can be seen as a rule for substituting either a path or a subcollection; a path can be seen as a subcollection by simply forgetting the order of the elements in the path. For example, the rule:

(x/x<3)+ as S → 3,4,5,():set

applied to the set 1,2,3,4,5,6,():set returns the set 3,4,5,6,():set because S matches the subset 1,2,():set and is replaced by the set 3,4,5,():set. The final result is computed as (3,4,5,():set)∪(3,4,5,6,():set).

Substitutions of Paths

Because the matched subcollection is also a path — that is, a sequence of elements — the seq type has a special role when appearing in the r.h.s. of a rule. If the r.h.s. evaluates to a sequence, and if this sequence has the same length as the matched path, then the first element of the sequence is used to replace the first element of the matched path, and so on, with the last element in the path replaced by the last element in the sequence. This convention is coherent with the subcollection substitution point of view and simplifies enormously the building of the r.h.s.

For example, suppose that in a GBF we want to model the random walk of a particle x. Then, two neighboring elements, one being x and the other undefined, must exchange their values. This is achieved with only one simple rule:

x,<undef>→<undef>,x

Figure 3. Random walk of a particle on the GBF <north, east>

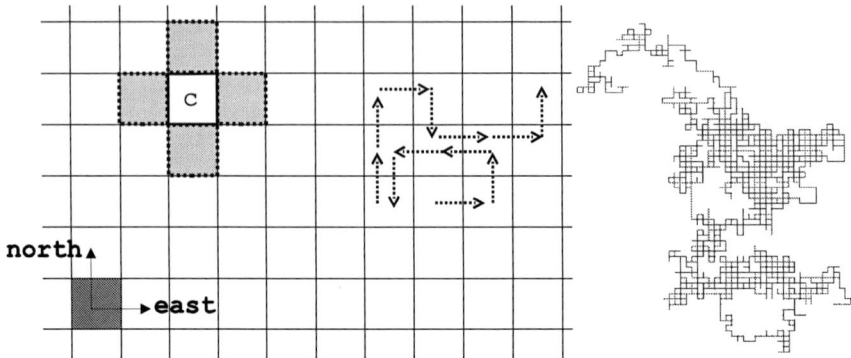

without the need to mention the precise neighborhood relationships between the two elements. Figure 3 illustrates this process on the GBF <north, east>. This free abelian GBF describes the usual rectangular lattice. Each cell c has four neighbors. Each application of the previous transformation moves the value in a cell to an empty neighbor cell. The path to the right of Figure 3 represents 3,000 random moves on this lattice.

We have mentioned above that the result of replacing a subset by a set is computed using set union. More generally, the insertion of a collection *C* in place of a subcollection *B* depends on the "borders" and on the topology of the involved collections. For example, in a sequence, the subcollection *B* defines in general two borders that are used to glue the ends of collection *C*. The gluing strategy may admit several variations. The programmer can select the appropriate behavior using the rule's attributes.

Transformations

A transformation R is a set of rules:

trans R = { ...; rule; ... }

For example, transformation

trans M = { x→x+1; }

defines a function M. The expression M(c) denotes the application of one transformation step to the collection c. A transformation step consists of the parallel application of the rules (modulo the rule application's features). So, M(c) computes a new collection c' where each element of c is incremented by one.

A transformation step can be easily iterated:

M[iter=n](c)

denotes the application of n transformation steps, and

M[iter=fixpoint](c)

denotes the application of M until a fixed point is reached; that is, the result c' satisfies the equation: c'=M(c').

PARADIGMS OF PATTERN FORMATION

In this section, we introduce several fundamental paradigms of pattern formation through some examples and their implementation in MGS. These examples are all fundamental models that have been proposed and discussed in the field of developmental biology. The purpose of this section is not to develop new developmental mechanisms, but to show that these paradigmatic examples can all be easily expressed in the unified framework of topological collection rewriting.

Diffusion and Beyond

Diffusion in a Continuous Setting
Just as thermal gradients cause heat to flow from a warmer area to a colder area, chemical gradients due to variations in chemical concentration cause molecules to move from high concentration to low concentration. This process, due solely to a concentration gradient, is referred to as *diffusion*. The rate of change in concentration with time and space is defined by Fick's law:

$$\frac{\partial C}{\partial t} = D \frac{\partial^2 C}{\partial z^2}$$

where C is the concentration of the chemical (mass/volume), D is some diffusion constant, and z is the spatial variable (to keep the example simple, we suppose that the diffusion occurs in a line).

A forward difference discretization gives

$$C(i,t+\Delta t)=(1-2h)C(i,t)+h\big(C(i-1,t)+C(i+1,t)\big)$$

where $C(i, t)$ represents the concentration at time t of the element of length i. Parameter h depends on the chemical and on the diffusion constant and must be less than 0.5. For the boundary conditions, we assume a source of constant concentration $C(0, t) = C_0$ at one end and a sink that ensures a constant concentration $C(n, t) = 0$ at the other end.

This very simple model can be programmed in MGS in the following way. We first have to define a sequence of elements representing a concentration in a line. This is simply a GBF with only one generator:

```
gbf Line = <right, left; right+left = 0>
```

The generator left is simply an alias for the inverse of right. These names can be used in a transformation to access the left and right neighbors of an element matched by a pattern variable in a Line. The position (a group element) of the element matched by a pattern variable x is simply denoted by pos(x). Beware that pos, left and right are not functions but special forms that have a meaning only within a transformation. These forms accept as argument only a pattern variable referring to one element in the collection. The transformation that makes the concentration to evolve can be simply written as:

```
trans diffuse[h, C0, n] = {
    x / pos(x) == (0*|right>) → C0;
    x / pos(x) == ((n-1)*|right>) → 0;
    x → (1-2*h)*x + h*(right(x) + left(x))
}
```

h, C0 and n are additional parameters of the transformation. The first two rules deal with the boundary conditions. We assume that the first element of the discretized line is put at the 0*|right> position (this denotes the identity element in the group of positions). Then, the last element is at position (n-1)*|right>. By default, the rules are applied with a priority corresponding to their order of

Figure 4. Result of the diffusion with parameter h=0.156, n=10,C0=30 and two different boundary conditions

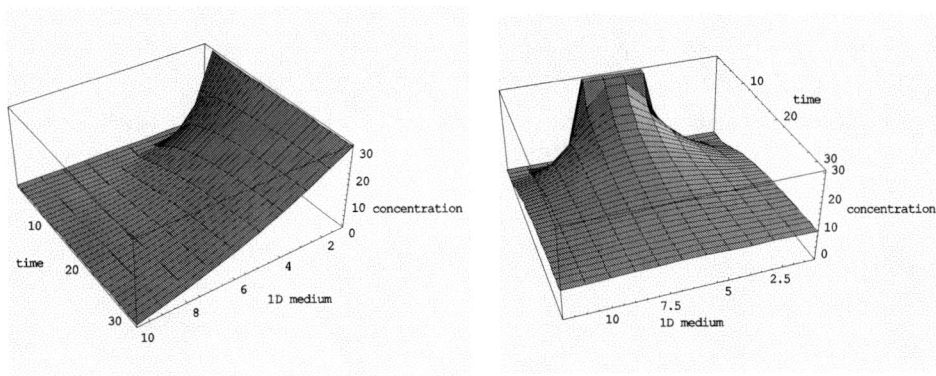

declaration: the first rule is applied whenever it can, then the second rule, and the third is possibly applied on the remaining subcollection. The two first rules specifying the behavior on the boundary take precedence to the last rule that governs the default behavior of the interior points.

The last step is to set the initial state S0 of the line. The operator |right> can be used to build this initial collection:

S0:=C0 |right> 0 |right> ... |right> 0

The evolution from 100 steps is then evaluated by the expression

diffuse[h, C0, n, iter=100](S0)

The results are visualized in Figure 4. The diagram at the left illustrates the diffusion process with parameter h=0.156, n=10,C0=30,and for 90 time step. The same process but with a different initial state and boundary conditions (no source or sinc) is illustrated on the right.

Diffusion at the Molecular Level

The previous example is very simple but still shows the ability of MGS to handle a continuous model. It is easy to extend this process to a surface or a volume instead of a line.

We now want to take the same system but focus on the level of molecules. The line is still discretized as a sequence of small boxes, indexed by a natural integer, and each containing zero or many molecules. At each time step, a molecule can choose to stay in the same box or to jump to a neighboring box with the same probability (see Figure 5). The state of a molecule is the index of the box in which it resides. The entire state of the system is then represented as a multiset of indices. The evolution of the system can then be specified as a transformation with three rules:

```
trans diffuseM = {
     n → n-1
     n → n
     n → n+1
}
```

Specifically for this example, the rules are applied non-deterministically with the same equal probability. The first rule specifies the behavior of a particle that jumps to the box at the left, the second rule corresponds to a particle that stays in the same box, and the last rule defines a particle jumping to the right.

Figure 5. Right diagram: principle of the particle diffusion model. Left diagram: result of a simulation

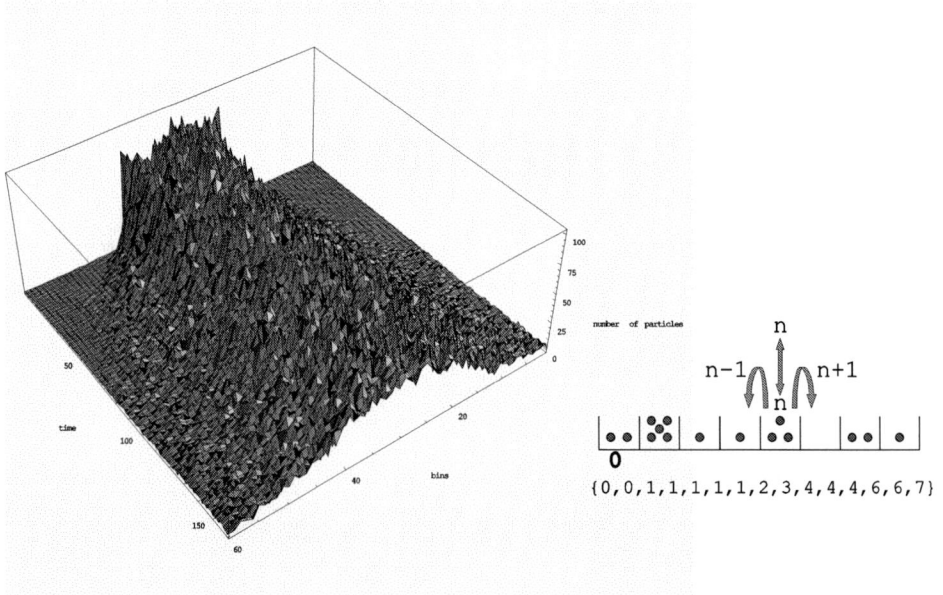

Figure 5 illustrates this approach and plots the result of the discrete diffusion of 1,500 particles on a sequence of 60 bins during 160 time steps. The diffusion is limited on the boundary (no flow, which is achieved by adding two additional rules to handle the behavior at the boundaries). In the initial states, all particles are randomly distributed in the middle third of the linear media (compare with the right side of Figure 4).

This example shows that even if a multiset has very little organization, it can be used to take geometric information into account. This model can also be extended to diffusion in a surface or a volume, and for arbitrary geometry. The idea is to discretize the medium in a set of bins and to represent the state of the system as a multiset of bins.

Boundary Growth

In this subsection, we focus on variations of cellular automata (von Neumann, 1966) used to model several growth processes.

The Growth of a Snowflake

This cellular automaton idealizes the formation of a snowflake (Wolfram, 2002). Black cells represent regions of solid ice, and white cells represent regions of liquid or gas. The molecules in a snowflake lie on a simple hexagonal grid. Whenever a piece of ice is added to the snowflake, a little heat is released, which then tends to inhibit the addition of further pieces of ice nearby. The corresponding evolution rule is very simple: a cell becomes black whenever exactly one of its neighbors was black in the previous step.

```
Trans Snowflake = {
    0 as x / neighborsfold(+, 0, x)==1 → 1
}
```

A black cell has the value 1 and a white cell has the value 0. A 0 is turned into a 1 only if the sum of its neighbors is one. The sum of the neighbors is computed using the **neighborsfold** operator that iterates an accumulating function over the neighbors:

$$neighborsfold(f, \ zero, \ x)$$

computes

$$f(x_1, f(x_2, \dots , f(x_n, zero)\dots)$$

Figure 6. The growth of a snowflake

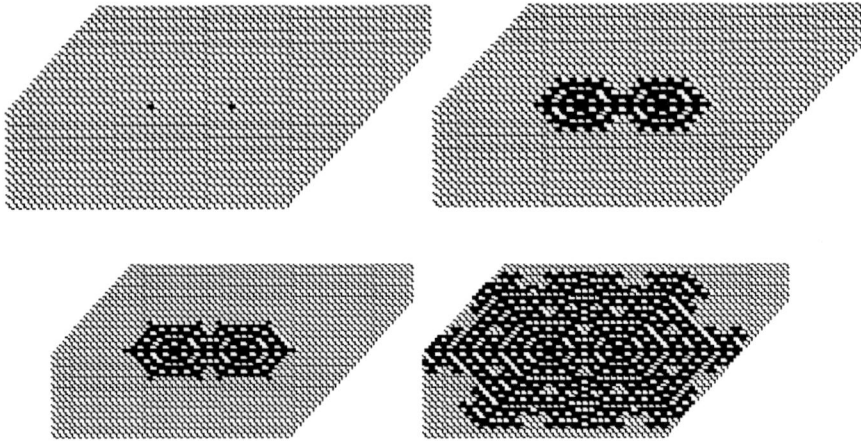

where the x_i are the neighbors of the element x. This operator can be used only within a transformation, and its last argument must be a pattern variable introduced in the left-hand side of the rule. Four steps of the evolution are pictured in Figure 6 (in the initial state, two black cells represent the ice).

Diffusion-Limited Aggregation

Diffusion-Limited Aggregation, or DLA, is a simple process of cluster formation by particles diffusing through a medium that jostles the particles as they move. When particles have the possibility to attract each other and stick together, they may form aggregates. The aggregates may grow as long as there are particles moving around. During the diffusion of a particle it is more likely that it attaches to the outer regions than to the inner ones of the cluster. Thus, a fractal shape occurs like that of corals or trees.

In the following implementation, a value 0 means a diffusing particle, while a value n greater than 1 means a particle fixed to a static cluster for n time steps. Then, the corresponding transformation is straightforward:

Figure 7. A DLA growing process on an hexagonal grid

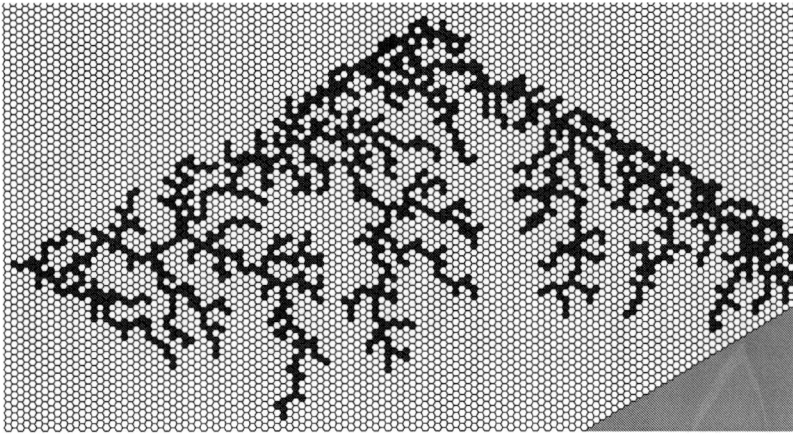

```
trans DLA = {
    aggregation =  0,n/n>0 → 1,n
    diffusion =    0,<undef> → <undef>,0
    timecount =    n / n > 0 → n+1
}
```

The rule **aggregation** specifies that a moving particle that comes near to a cluster will become part of this cluster. The rule **diffusion** defines the random move of a particle: the particle occupies a free neighboring cell and empties its current occupied cell. The last rule updates the age of a particle stuck into a cluster. The result is illustrated in Figure 7. In this simulation, particles are constantly created on the right, and the initial static particle is completely on the left. Particles are constrained to move on a bounded rhomboidal domain. This explains the asymmetries of the figure.

Phenomenological Sketch of the Growth of a Tumor

This model illustrates the growth of a tumor. It is inspired from a model initially proposed in Wilensky (1998). We start by modeling a set of cells and the mechanical forces between them. Then we add a growing process by giving two different behaviors to the cells. This model gives a formal example of an interacting set of entities localized in a 3-D space, such that the interactions both depend on the position of the entities and make these positions evolve.

The Mechanical Model of the Cells

Each cell exerts a spring-like force to its neighbors. The resulting forces induce cell movements. To keep the model simple, we assume an Aristotelian mechanical physics; that is, the velocity of the cell is proportional to the force exercised. Although this is not compatible with Newtonian physics (the acceleration is actually proportional to the force), the final state (the position of the cells at the equilibrium) is the same and avoids the handling of the acceleration variables.

We represent each cell as a point in $\mathbf{R}^3$ (e.g., assimilated to its mass center) and with some velocity:

```
record Point = {x, y, z}
record Cell = Point + {vx, vy, vz, l, age}
```

These two statements define two new record types named Point and Cell. A Point is a record with the fields x, y, and z (for recording the position of a cell). The Cell type is a subtype of Point having, in addition, the fields vx, vy, and vz (for recording the velocity of a cell) and the fields l and age that record the radius and the age of the cell.

The interaction force between two cells is computed by the function interaction:

```
fun interaction(ref, src) =
    let X = ref.x ▯ src.x
    and Y = ref.y ▯ src.y
    and Z = ref.z ▯ src.z
    and L = ref.l + src.l
    in let dist = sqrt(X*X + Y*Y + Z*Z)
    in let force = (L - dist)/dist
    in {fx=X*force, fy=Y*force, fz=Z*force}
```

The result is a record with fields fx, fy, and fz, which represents the coordinates of the force vector exercised on the cell ref by the cell src. This force goes to infinity when the two interacting cells become closer, it vanishes when the cells are separated by their natural diameter, and becomes asymptotically proportional to the distance between the cells when this distance increases.

A transformation is used to iterate over the cell and to compute the resulting forces:

```
trans Meca = {
    e:Cell  →
    let tf = neighborsfold(sum(e), {fx=0,fy=0,fz=0}, e)
    in e + { x  = e.x + epsilon*e.vx,
             y  = e.y + epsilon*e.vy,
             z  = e.z + epsilon*e.vz,
             vx = tf.fx,
             vy = tf.fy,
             vz = tf.fz }
}
```

The pattern e:Cell is equivalent to e/Cell(e) and selects one element of type Cell. The operator + that appears in the body of the let is overloaded. In addition to the standard numeric addition, it denotes the asymmetric merge of two records. If r and s are two records, then the record r+s contains all of the fields present in r and s. The value of the field f of the record r+s is the value of s.f if f exists in s and else r.f. The net effect of the expression in the body of the let is a record similar to e where the position and the velocity have been updated.

The operator neighborsfold iterates a binary function over the neighbors of an element to compute the total force tf exercised on the cell matched by e. Function sum(e) computes the interaction between e and a neighbor cell and accumulates the result:

```
fun sum(e,s,acc) = addv(acc, interaction(e, s))
fun addv(u,v) = {fx=u.fx+v.fx, fy=u.fy+v.fy, fz=u.fz+v.fz}
```

Note that the function sum is curryed and partially applied in the application of neighborsfold. The operator neighborsfold is similar to the *fold* in functional languages (where it iterates over lists or other algebraic data types) and the use of a curryfied function as the functional argument of the fold is a well-known and heavily used programming pattern (Sheard et al., 1993).

The neighborhood of a cell is computed dynamically using a Delaunay graph built from the cell positions: for a set S of points in the $\mathbf{R}^d$, the Delaunay graph is the unique triangulation of S such that no point in S is inside the circumcircle of any triangle. At each time step, this neighborhood can change due to the cell movements. In MGS, the Delaunay collection type is a type of constructor corresponding to the building of a collection with a neighborhood computed from the position of the elements in a d-dimensional space. A Delaunay collection type is specified by giving the function that extracts the sequence of coordinates from an element of the collection:

```
delaunay(3) D3 =
\p.if Point(p) then p.x, p.y, p.z
    else error(«Bad element type for a D3») fi
```

The parameter 3 after the keyword **delaunay** indicates that the elements of this collection type correspond to points in $\mathbf{R}^3$. The notation \x.e is the syntax used for the lambda-expression $\lambda x.e$.

Figure 7 illustrates the trajectory of seven cells computed by iterating the transformation **Meca** over a collection D3.

The Behavioral Model of the Cells
   A tumor consists of stem and transitory cells:

- A transitory cell moves, subject to the forces exercised by other cells.
- A transitory cell may divide at age **tc_d** with a probability **tc_p**.
- A transitory cell with an age greater than **tc_a** eventually dies with a probability **tc_d**.
- A stem cell is fixed (e.g. anchored in the extra-cellular matrix).

Figure 7. A trajectory of seven cells attracted by a spring-like force by the neighbors

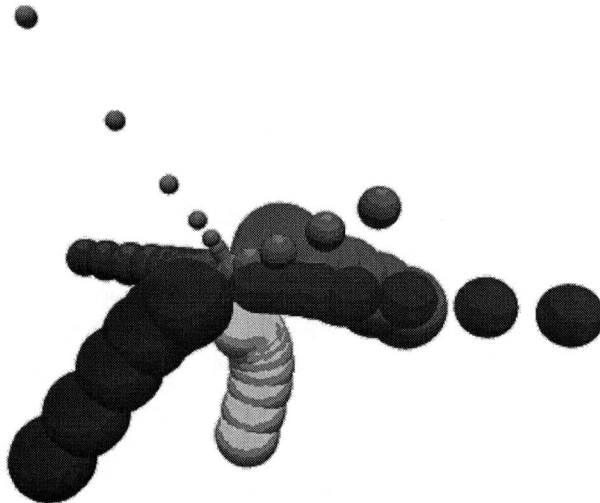

- A stem cell can divide either asymmetrically or symmetrically at some age sc_d. In either case, one of the two daughter cells remains a stem cell, replacing its parent. In asymmetric mitosis, the other daughter is a transitory cell. In symmetric mitosis, the other daughter is a stem cell that is allowed to move for sc_w time before anchoring and being static. The probability of an asymmetric mitosis is sc_p.

These behaviors are specified by the transformation Grow. To select the appropriate cell in the left-hand side of a rule, we introduce the types StemCell and TransitoryCell that are distinguished only by the presence (or the absence) of the field stem.

```
record StemCell = Cell + {stem=true}
record TransitoryCell = Cell + {~stem}
```

We represent a young stem cell allowed to move as a transitory cell with a negative age. When such kind of cells reaches the age -1, then they transform themselves into static stem cells. Then the definition of Grow can be:

```
trans Grow = {
    c:TransitoryCell / c.age == -1
    →      c+{stem=true, age=0};
    c:TransitoryCell / (c.age > tc_a) & (rand() < tc_d)
    →      <undef>;
    c:TransitoryCell / (c.age == tc_d) & (rand() < tc_p)
    →      r, r+{x=noise(c.x), y=noise(c.y), z=noise(c.z), age=0};
    c:StemCell / c.age == sc_d
    →      c, { x=noise(c.x), y=noise(c.y), z=noise(c.z),
                vx=0, vy=0, vz=0, l=c.l,
                age = if rand()<sc_p then -sc_w else 0 fi
           }
    c:Cell
    →      c+{age=c.age+1};
}
```

The pseudo-function rand returns a random number between 0 and 1. The function noise perturbs its argument: fun noise(x) = x + rand(). So when a cell divides, one of the two daughter cells inherits the position of the mother cell and the other is a little away. The mechanical interaction then moves the cell away. To restrict the mechanical effects to the transitory cells, it is enough to match

Figure 8. Growth of a tumor (See "Phenomenological Sketch of the Growth of a Tumor" for further explanations)

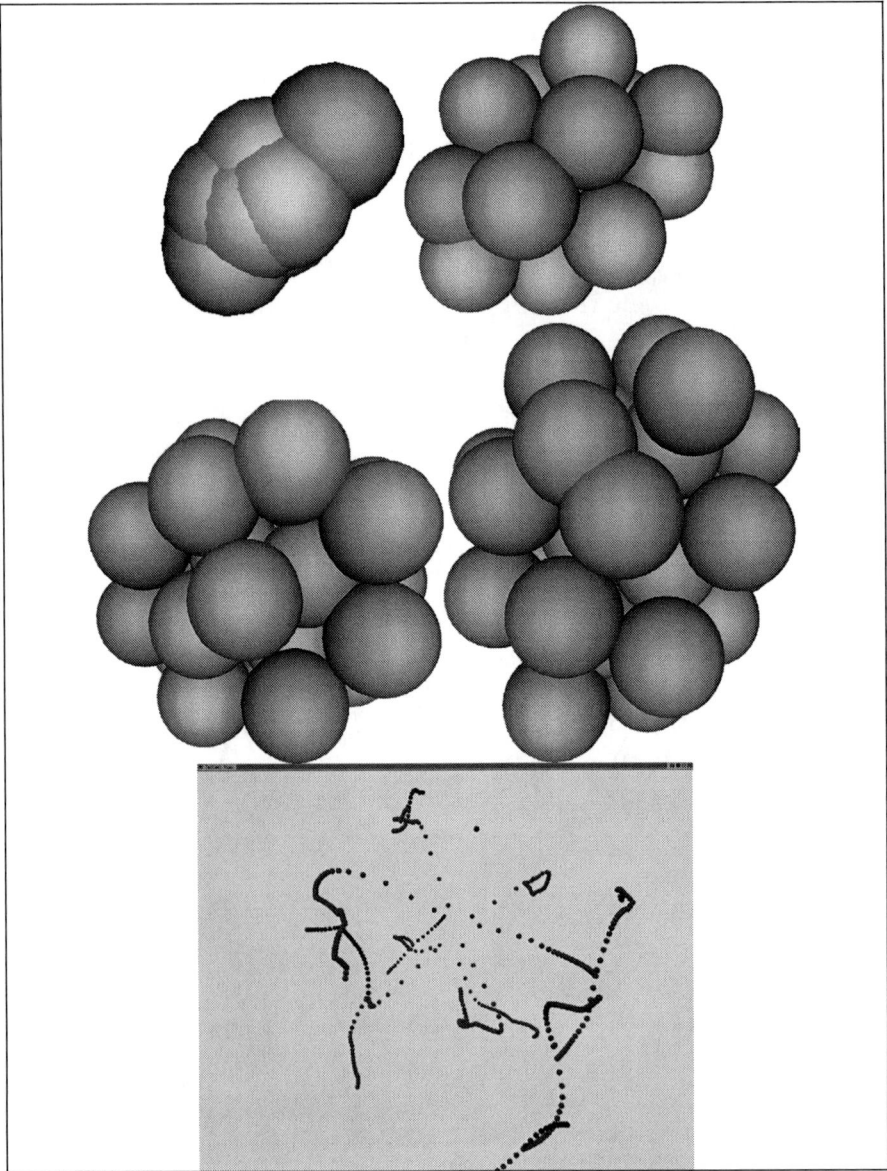

TransitoryCell instead of Cell in the transformation Meca: the left-hand side of the rule becomes e:TransitoryCell.

Figure 8 illustrates some states in the tumor progression. The first four images are drawings of the tumor at different time moments (cells have divided and rearranged to satisfy the mechanical constraints). The last image (at the bottom side) corresponds to the trajectory of the mass center of each cell, before the first mitosis of a stem cell. The division of a transitory cell (visible as forks in the trajectories) induces a change in the mechanical constraints and a corresponding change in the trajectory.

This kind of simulation is phenomenological because the behavior of each cell is roughly modeled, without taking into account the chemical and genetic processes. However, this kind of simulation can be used to estimate the propagation of the tumor, to evaluate the ratio between transitory and stem cells, and to evaluate the impact of various therapeutic strategies, such as cell division or mobility inhibitors. Most of the chemotherapy drugs known as M- and S-poisons inhibit cell division and impact mainly the transitory cells. The problem is that the stem cells (also known as clonogenic cells) maintain the tumor and propagate its metastases. Other possible approaches try to lower the cell mobility to reduce the tumor propagation.

DISCUSSION AND CONCLUSION

Summary

The MGS programming language is largely inspired by the dynamical system perspective on biological development. From this point of view, biological development exhibits a dynamical structure, or a variable phase space, that must be computed jointly with the current state of the system. While it makes the classical modeling and the simulation of such systems very difficult, their description is still usually easily achieved by a set of local evolution rules specifying the transformation of a subsystem. However, the partition of the system into subsystems evolves in the course of time and the conditions of a transformation application can be complex. These considerations lead to the development of a rule-based programming paradigm. This programming paradigm is characterized by the repeated, localized transformations of a shared data object. The transformations are described by *rules* that separate the description of the subobject to be replaced (the *pattern*) from the calculation of the replacement.

Optionally, rules can have further conditions that restrict their applicability and the transformations are controlled by explicit or implicit *strategies*. When the data object is a term, we retrieve the notion of rewrite systems. MGS extends this approach by considering objects structured by neighborhood relationships. The topological approach unifies several models of computations, at least to provide a single syntax that can be consistently used to allow the merging of these formalisms for programming purposes. A unifying theoretical framework can be developed (Giavitto & Michel, 2001; Giavitto & Michel, 2002), based on the notion of *chain complex* developed in algebraic topology.

The resulting programming style is an effective framework for the modeling and the simulation of various developmental processes, as shown in the previous section. All of the examples in this chapter have been processed using the MGS interpreter. Theoretical articles, documentations, and various MGS software products are freely available at: *http://mgs.lami.univ-evry.fr*.

Related Works

Transformation on multisets is reminiscent of multiset rewriting (or rewriting of terms modulo associativity and commutativity). This is the main computational device of Gamma (Banâtre & Le Metayer, 1986; Banâtre et al., 1987, 2001). The CHemical Abstract Machine (CHAM) extends these ideas with a focus on the semantics of nondeterministic processes (Berry & Boudol, 1992). The CHAM introduces a mechanism to isolate some parts of the chemical solution. This idea has been seriously taken into account in the context of P systems. P systems (Păun, 2001) are a new distributed parallel computing model based on the notion of a membrane structure. A membrane structure is a nesting of cells represented by a Venn diagram without intersection and with a unique superset: the skin. Objects are placed in the regions defined by the membranes and evolve following various transformations: an object can evolve into another object, can pass through a membrane, or dissolve its enclosing membrane. As for Gamma, the computation is finished when no object can further evolve. By using nested multisets, MGS is able to emulate more or less the notion of P systems. In addition, patterns like the iteration + go beyond what is possible to specify in the l.h.s. of a Gamma rule.

Lindenmayer systems (Lindenmayer, 1968) have long been used in the modeling of $(DS)^2$ (especially in the modeling of plant growth). They loosely correspond to transformations on sequences or string rewriting (they also correspond to tree rewriting because some standard features make arbitrary trees that are particularly simple to code). Obviously, L systems are dedicated to the handling of linear and tree-like structures.

Strong links exist between GBF and cellular automata (CA), especially considering the work of Z. Róka, which has studied CA on Cayley graphs (1994). However, our own work focuses on the construction of Cayley graphs as the shape of a data structure, and we have developed an operator algebra and rewriting notions on this new data type. This is not in the line of Z. Róka's works, which focuses on synchronization problems and establishes complexity results in the framework of CA.

In the domain of biological process modeling, an increasing research effort is devoted to the design of a simulation platform at the cellular level. The current projects are mainly based on the modeling of the metabolic activities through differential equations (ODEs) or partial differential equations (PDEs). They act then as ODE or PDE solvers dedicated to biological processes. For instance, BioDrive (Kyoda et al., 2000) handles signal transduction. E-Cell is dedicated to the design of a minimal set of genes coding basic metabolic functions and includes an evaluation of the corresponding cell functioning through an energetic cost (Tomita et al., 1999; *www.e-cell.org*). V-Cell is one of the most advanced simulation platforms — it handles PDEs and enables the specification of complex geometry (Schaff & Loew, 1999; *www.nrcam.uchc.edu*). These simulators rely on the hypothesis that the chemical activities that occur in the cells are adequately described only by their kinetic equations (as it is the case in a test tube). Consequently, they are unable to model and ignore the dynamic organizations that modify profoundly the reactions, such as compartimentalization, the creation of hyperstructures (Amar et al., 2002), or the channeling in the case of metabolon by the so-called "solid-state metabolism" effect.

Other modeling approaches rely on the multi-agent paradigm to model the various entities and activities that appear in biological processes. The resulting software architecture support fits very well into the description of the domain's ontology (the specification of an ontology of biological entities and activities is a problem in itself considering the vast number of different entities and activities to describe). However, this approach does not bring a solution by itself to the problem of describing the interaction between an arbitrary collection of agents and the spatial organization of the agents. To overcome this problem, some current projects extend the multi-agent paradigm with other approaches, such as cellular automata.

The MGS programming style corresponds to the rule-based programming paradigm. Rule-based programming is currently experiencing a renewed period of growth with the emergence of new concepts and systems that allow a better understanding and better usability. However, the vast majority of rule-based languages (such as expert systems) are funded on a logical approach (computation is a logical deduction), which is not adequate to describe various biochemical processes. Yet the algorithmic and biological examples given in the two previous sections demonstrate the ability of MGS to express both discrete and (the numerical solution of) continuous models.

Perspectives

The perspectives opened by this preliminary work are numerous. We want to develop several complementary approaches to define new topological collection types. One approach to extend GBF applicability is to consider monoids instead of groups, especially automatic monoids, which exhibit good algorithmic properties. Another direction is to handle general combinatorial spatial structures such as simplicial complexes. At the language level, the study of the topological collections concepts must continue with a finer study of transformation types. Several kinds of restrictions can be considered with regard to these transformations, leading to various kinds of pattern languages and rules. The complexity of matching such patterns has to be investigated. The efficient compilation of an MGS program is a long-term research plan. We have considered in this chapter only one-dimensional paths, but a general n-dimensional notion of paths exists and can be used to generalize the substitution mechanisms of MGS. From the applications point of view, we are looking for simulations of more complex developmental processes.

ACKNOWLEDGMENTS

The authors would like to thank M. Gheorghe at the University of Sheffield, R. Paton and G. Malcolm at the University of Liverpool, F. Delaplace at the University of Evry, C. Godin and P. Barbier de Reuille at CIRAD-Montpellier, the members of the epigenomic group at GENOPOLE-Evry, P. Prusinkiewicz at the University of Calgary, and the organizers of the friendly "workshop on membrane computing" series for helpful discussions, biological motivations, fruitful examples, and challenging questions. Further acknowledgments are also due to J. Cohen, A. Spicher, B. Calvez, F. Thonnerieux, C. Kodrnja, and F. Letierce who have contributed in various ways to the MGS software. This research is supported in part by the French National Center for

Scientific Research (CNRS), GENOPOLE-Evry, the national working group GDR ALP and IMPG, and the University of Evry Val d'Essonne.

REFERENCES

Amar, P., Giavitto, J.-L., Michel, O., Norris, V., & 36 other co-authors (2002). Hyperstructures, genome analysis and I-cells. *Acta Biotheoretica, 50*(4), 357-373.

Banâtre, J.-P. & Le Metayer, D. (1986). A new computational model and its discipline of programming. *Technical Report RR-0566*, Inria.

Banâtre, J.-P., Coutant, A., & Le Metayer, D. (1987). Parallel machines for multiset transformation and their programming style. *Technical Report RR-0759*, Inria.

Banâtre, J.-P., Fradet, P., & Le Metayer, D. (2001). Gamma and the chemical reaction model: Fifteen years after. In Calude, C.S., Păun, Gh., Rozenberg, G., & Salomaa, A. (Eds.), *Multiset processing: Mathematical, computer science, and molecular computing points of view. Lecture Notes in Computer Science*, 2235, Berlin: Springer, 17-31.

Berry, G. & Boudol, G. (1992). The chemical abstract machine. *Theoretical Computer Science, 96*, 217-248.

Bournez, O., Kirchner, H., Côme, G.-M., Conraud, V., & Ibanescu, L. (2003). A rule-based approach for automated generation of kinetic chemical mechanisms. In Nieuwenhuis, R. (Ed.), *14th Int. Conf. on Rewriting Techniques and Applications. Lecture Notes in Computer Science*, 2706, Berlin: Springer, 30-45.

Buneman, P., Naqvi, S., Val Tannen, B., & Wong, L. (1995). Principles of programming with complex objects and collection types. *Theoretical Computer Science, 149* (1), 3-48.

Dershowitz, N. (1993). A taste of rewrite systems. Lecture notes in *Computer Science*, 693, Berlin: Springer, 199-228.

Dershowitz, N. & Jouannaud, J.-P. (1990). Rewrite systems. *Handbook of Theoretical Computer Science*, Vol. B. 244-320.

Dittrich, P., Ziegler, J., & Banzhaf, W. (2001). Artificial chemistries – a review. *Artificial Life, 7* (3), 225-275.

Eigen, M. & Schuster, P. (1979). *The hypercycle: A principle of natural self-organization.* Berlin: Springer.

Eker, S., Knapp, M., Laderoute, K., Lincoln, P., Meseguer, J., & Kemal Sönmez, M. (2002). Pathway logic: Symbolic analysis of biological signaling. In *Pacific Symposium on Biocomputing PSB 2002*, 400-412.

Fisher, M., Malcolm, G., & Paton, R. (2000). Spatio-logical processes in intracellular signalling. *BioSystems*, 55, 83-92.

Fontana, W. & Buss, L. (1994). "The arrival of the fittest": Toward a theory of biological organization. *Bulletin of Mathematical Biology*, 56, 1-64.

Giavitto, J.-L. (2001). Declarative definition of group indexed data structures and approximation of their domains. In *ACM SIGPLAN Conference on Principles and Practice of Declarative Programming PPDP01*. ACM Press, 150-161.

Giavitto, J.-L. (2003). Topological collections, transformations and their application to the modeling and the simulation of dynamical systems. In Nieuwenhuis, R. (Ed.), *14th International Conference on Rewriting Techniques and Applications RTA'03. Lecture Notes in Computer Science*, 2706, Berlin: Springer, 208-233.

Giavitto, J.-L. & Michel, O. (2001). MGS: A programming language for the transformations of topological collections. *LaMI Technical Report 61-2001*. University of Évry Val d'Essonne, France.

Giavitto, J.-L. & Michel, O. (2002). The topological structures of membrane computing. *Fundamenta Informaticae*, 49 (1-3), 107-129.

Giavitto, J.-L. & Michel, O. (2003). Modeling the topological organization of cellular processes. *BioSystems*, 70(2), 149-163.

Giavitto, J.-L., Malcolm, G., & Michel, O. (2004). Rewriting systems and the modelling of biological systems. *Comparative and Functional Genomics*, 5(1), 95-99.

Giavitto, J.-L., Michel, O., & Sansonnet, J.-P. (1995). Group based fields. In *Parallel Symbolic Languages and Systems PSLS95. Lecture Notes in Computer Science*, 1068, Berlin: Springer, 209-215.

Hammel, M. & Prusinkiewicz, P. (1996). Visualization of developmental processes by extrusion in space-time. *Proceedings of Graphics Interface '96*, 246-258.

Harper, J.L., Rosen, B.R., & White, J. (1986). *The growth and form of modular organism*. London: The Royal Society.

Head, T. (1987). Formal language theory and DNA: An analysis of the generative capacity of specific recombinant behaviors. *Bulletin of Mathematical Biology*, 49 (6), 737-759.

Head, T. (1992). Splicing schemes and DNA. In *Lindenmayer Systems: Impacts on Theoretical Computer Science, Computer Graphics, and Developmental Biology*. Berlin: Springer, 371-383. Also appears in (1992). *Nanobiology*, 1, 335-342.

Hung, J.Y., Gao, W., & Hung, J.C. (1993). Variable structure control: A survey. *IEEE Transactions on Industrial Electronics*, *40* (1), 2-22.

Itkis, Y. (1976). *Control Systems of Variable Structure*. New York: Wiley.

Kaufman, S. (1995). *The origins of order: Self-organization and selection in evolution*. Oxford: Oxford University Press.

Keller, E.F. (1995). *Refiguring life: Metaphors of twentieth-century biology*. New York: Columbia University Press.

Kephart, J. & Chess., D. (2003). The vision of autonomic computing. *IEEE Computer Magazine*, *36*(1) 41-50.

Kyoda, K.M., Muraki, M., & Kitano, H. (2000). Construction of a generalized simulator for multi-cellular organisms and its application to Smad signal transduction. In *Fifth Pacific Symposium on Biocomputing PSB 2000*, 314-325.

Lincoln, P. (2003). Symbolic systems biology. In Nieuwenhuis, R. ed. *14th International Conference on Rewriting Techniques and Applications RTA'03. Lecture Notes in Computer Science*, 2706, Berlin: Springer, 1.

Lindenmayer, A. (1968). Mathematical models for cellular interaction in development, parts I and II. *Journal of Theoretical Biology*, 18, 280-299 and 300-315.

Maynard-Smith, J. (1999). *Shaping life: Genes, embryos and evolution*. New Haven, CT: Yale University Press.

Meinhardt, H. (1982). *Models of biological pattern formation*. New York: Academic Press.

Michel, O. (1996). Design and implementation of 81/2, a declarative data-parallel language. *Computer Language*, *22*(2/3), 165-179.

Mjolsness, E., Sharp, D.H., & Reinitz, J. (1991). A connectionist model of development. *Journal of Theoretical Biology*, *152* (4), 429-454.

Munkres, J.R. (1993). *Elements of algebraic topology*. Addison-Wesley.

Parashar, M. & Hariri, S., eds. (2003). Autonomic applications workshop. Taj Krishna, Hyderabad, India. Special issue of *Cluster Computing, The Journal of Networks, Software Tools and Applications* (2004).

Paton, R. ed. (1994). *Computing with biological metaphors*. London, New York: Chapman & Hall.

Păun, Gh. (2001). From cells to computers: Computing with membranes (P systems). *BioSystems*, *59*(3), 139-58.

Prusinkiewicz, P. (1998). Modeling of spatial structure and development of plants: A review. *Scientia Horticulturae*, 74, 113-149.

Prusinkiewicz, P. (1999). A look at the visual modeling of plants using L-systems. *Agronomie*, 19, 211-224.

Prusinkiewicz, P. & Lindenmayer, A. (1990). *The algorithmic beauty of plants*. Berlin: Springer.

Róka, Z. (1994). One-way cellular automata on Cayley graphs. *Theoretical Computer Science*, *132* (1–2), 259-290.

Schaff, J. & Loew, L.M. (1999). The virtual cell. In *Fourth Pacific Symposium on Biocomputing PSB 1999*, 4, 228-239.

Sheard, T. & Fegaras, L. (1993). A fold for all seasons. *Proceedings of the 6th ACM SIGPLAN/SIGARCH International Conference on Functional Programming Languages and Computer Architecture FPCA'93*, ACM Press, 233-242.

Stengers, I. (1988). D'une science à l'autre. *Les Concepts Nomades*. Paris: Le Seuil.

Stevens, P.S. (1974). *Patterns in nature*. Boston: Little, Brown and Co.

Thompson, D'Arcy W. (1942). *On growth and form*. Cambridge: University Press.

Tomita, M., Hashimoto, M., Takahashi, M., Shimizu, T.-S., Matsuzaki, Y., Miyoshi, F., Saito, K., Tanida, S., Yugi, K., Venter, J.-C., & Hutchison, C.-A., (1999). E-CELL: Software environment for whole cell simulation. *Bioinformatics*, *15* (1),72-84.

Turing, A.M. (1952). The chemical basis of morphogenesis. *Philosophical Transactions*. Royal Society of London, Series B: Biological Sciences, (237), 37-72.

Varela, F.J. (1979). *Principle of biological autonomy*. New York: McGraw-Hill/Appleton and Lange.

von Neumann, J. (1966). *Theory of self-reproducing automata*. Urbana, IL: University of Illinois Press.

Wilcox, M., Mitchison, G.J., & Smith, R.J. (1973). Pattern formation in the blue-green alga, *Anabaena*. I. Basic mechanisms. *Journal of Cell Science*, 12, 707-723.

Wilensky, U. (1998). NetLogo tumor model. Contributed by Gershom Zajicek M.D., Prof. of Experimental Medicine and Cancer Research at The Hebrew University-Hadassah Medical School, Jerusalem. *http://ccl.northwestern.edu/netlogo/models/Tumor*. Center for Connected Learning and Computer-Based Modeling, Northwestern University, Evanston, IL.

Wolfram, S. (2002). *A new kind of science*. Champaign, IL: Wolfram Media, Inc.

Wolpert, L., Beddington, R., Lawrence, P., Meyerowitz, E., Smith, J., & Jessell, T.M. (2002). *Principles of development* (2nd ed.). Oxford: Oxford University Press.

ENDNOTES

[1]   As stated by d'Arcy W. Thompson (1942): "We might call the form of an organism an event in space-time, and not merely a configuration in space."

Chapter VII

Computing Bacterial Evolvability Using Individual-Based Models

Richard Gregory, University of Liverpool, UK

Costas Vlachos, University of Liverpool, UK

Ray C. Paton, University of Liverpool, UK

John W. Palmer, University of Liverpool, UK

Q. H. Wu, University of Liverpool, UK

Jon R. Saunders, University of Liverpool, UK

ABSTRACT

This chapter describes two approaches to individual-based modelling that are based on bacterial evolution and bacterial ecologies. Some history of the individual-based modelling approach is presented and contrasted to traditional methods. Two related models of bacterial evolution are then discussed in some detail. The first model consists of populations of bacterial cells, each bacterial cell containing a genome and many gene products derived from the genome. The genomes themselves are slowly mutated over time. As a result, this model contains multiple time scales and is very fine-grained. The second model employs a coarser-grained, agent-based architecture designed to explore the evolvability of adaptive behavioural strategies in artificial bacterial ecologies. The organisms in this approach are represented by mutating learning classifier systems. Finally, the subject of computability on parallel machines and clusters is applied to these models, with the aim of making them efficiently scalable to the point of being biologically realistic by containing sufficient numbers of complex individuals.

INTRODUCTION

A systemic approach to the study of biological systems has been pursued for many decades, with its modern roots in a number of developing areas of nineteenth century science such as physiology (homeostasis, histology), tissues and systems, and ecology (lake as a microcosm). In the twentieth century we see major integrative developments, especially with the rise of General Systems Theory and cybernetics. More recently, with the increase in data available from molecular biological experiments, it has become important to describe an emerging field of systems biology. We wish to place our work within this long tradition of interdisciplinary science.

The approach to modelling biological systems that is presented here includes evolutionary and ecological perspectives. Coupled with this is an approach from the "bottom up" that seeks to model individual entities and processes as individuals rather than averaged aggregates. We shall explore some ways in which an individual-based approach that seeks to include ecological and evolutionary dimensions can be implemented by describing two systems we have developed that model (at different levels of biological granularity) bacterial systems in simple environments. Some of the many computational challenges to these approaches are described, as well as the biological realism we wish to incorporate in the models. A further challenge, and opportunity, is the increasing access to very large computational power. We discuss a number of aspects for using grid-enabled systems.

Individual-based modelling (IbM) provides an important complementary approach to biosystems modelling that relies on population-based (averaging) techniques (DeAngelis & Gross, 1992). Levins (1984) made the very important comment that no model can simultaneously optimise for generality, realism, and precision. There is always a trade-off between at least two of these modelling dimensions. Whereas some model developers may wish to provide general (though usually unrealistic) models, we seek to develop bottom-up models of complex evolving systems that preserve a degree of biological realism. To achieve this goal, we have developed a number of individual-based environments that allow biologists, from biochemists to ecologists, to explore computational models of a range of systems under study.

Modelling Biological Systems from the Bottom Up

A number of groups have approached IbMs from a computational stance that makes use of such computational architectures as abstract machines (automata), (quasi-)autonomous agents (distributed AI systems), and ALife. For example, the ecological simulators Echo (Hraber et al., 1997) and Herby (Devine & Paton, 1997) employed rule-based, AI-inspired architectures to explore the evolvability of ecosystems. Ginot et al. (2002) discuss a component programming approach using a multi-agent architecture that could help end users (e.g., with no programming expertise) to develop an individual-based modelling capability based on a set of primitive behavioural activities (such as *locate*, *select*, *translate*) and workflow control. An example of the practical utility of the IbM approach is reported by van Nes et al. (2003), who used a spatially explicit model (Charisma) to simulate the growth of competing species of submerged lake macrophytes. They note that although the behaviour of the model easily became complex and puzzling if more than one species were investigated, the approach allowed a "realistic, yet transparent" model.

There are many advantages of IbMs ranging from the greater option for realism to the lack of need (especially in computational models) to avoid difficult mathematics. However, there are serious caveats in these (as in any other) modelling systems. There is a danger in not only conflating material (matter), energy, and information in artificial (virtual) simulation systems, but also in creating things from nothing. This "economic" currency-balancing requirement is critical to any approach to systems biology that needs to deal with space and time, energy and material balances, and evolutionary processes. With regard to contemporary studies of "agents" found in many distributed AI systems, our work has never fit exactly into the more common rational or logical agent-based approach typified by such architectures as the DBI (desire, belief, intention) model. We were always motivated to constrain certain assumptions in these latter models, especially their strong reliance on anthropomorphism and teleology. To be fair, we have not escaped the need to require teleonomic dimensions related to the determination of fitness and adaptability.

With regard to bacterial IbMs (other than our own), we note the important system of Kreft et al. (1998) called BacSim. BacSim was inspired by the Echo system (Hraber et al., 1997). A recent development of BacSim (Kreft et al., 2001) simulates a multisubstrate, multispecies model of nitrifying biofilms. Biofilm growth involves diffusion, reaction, and growth, and the spread of biomass occurs by shoving cells to minimize their overlap and is governed by cellular automata rules. Biomass is distributed in a discrete grid and each species has uniform growth parameters. Another recent bacterial IbM system is INDISIM (Ginovart et al., 2002a). This is a discrete, stochastic simulator for studying bacterial cultures. It simulates the behaviour of the system in which its global properties emerge from the rule-following behaviour of individual bacteria. These rules require the input of only a few parameters. The state of each bacterium is determined by a set of random, time-dependent variables related to spatial location, biomass, and other individual properties. These variables may include concentrations of different types of particles, nutrients, reaction products, and residual products. Random variables are used to characterize the individual bacterium and the individual particle, as well as the updating of individual rules. Ginovart et al. (2002b) used INDISIM to study the influence of subspecies of *Streptococcus salivarius* and *Lactobacillus delbrueckii* in yogurt processing.

Our initial work was based on ecological systems and included ElVis and Herby (Devine et al., 1996). The former is an adaptive, agent-based ecological vision system that uses a fuzzy learning classifier system architecture. ElVis is a simulation system that is required to recognise gaps between trees as it navigates through an artificial woodland comprising randomly placed trees. No prior "knowledge" was encoded into this system except for recognition of vertical lines. Thus, two laterally adjacent vertical lines could be a gap or a tree. Rules were evolved or engineered to enable the system to process simulated visual inputs, and based on this acquired information, the system is better equipped to proceed between obstacles (Kendall, 1998). Herby is an artificial ecology consisting of herbivores and a heterogeneous patchy environment of different (and changing) food resources (Devine & Paton, 1997). In this system the agents have a few behavioural options — movement (in one of four general directions), reproduction, feeding, or doing nothing. There is always a differential material/energetic cost to actions by agents in the system, as well as an impact by them on their environment.

At the same time that we were developing individual-based models of ecological systems, we were also looking at individual-based models at the molecular and cellular levels. It had been recognised that the ecological metaphor was a valuable source idea for intracellular systems (e.g., Welch, 1987). This idea was more broadly expressed in the context of cellular information processing (e.g., Bray, 1995; Conrad, 1992; Paton, 1993). We started to describe individual molecular species in terms of agents with regard to signalling proteins including CdK and CaM Kinase II (Paton et al., 1996). This notion was developed even further with the proposal that at suitable levels of abstractness, objects (agents) and processes (actions) can be treated as the same kind of thing (Paton & Matsuno, 1998). Many proteins, such as enzymes and transcription factors, operate as context-sensitive information processing agents and display a number of "cognitive" capacities including pattern recognition, fuzzy data handling, memory capacity, multifunctionality, signal amplification, integration, and crosstalk (Paton et al., 1996; Fisher et al., 1999).

There has also been a shift in the perception of hereditary material towards a more "cognitive" stance. Indeed, a number of metaphors interplay in contemporary discussions on characterising hereditary information (Avise, 2001; Keller, 1995; Konopka, 2002; Paton, 1997). Table 1 contrasts the more mechanical-chemical view of hereditary material with a contemporary, information-rich, fluid genome. Within a contemporary framework, genomes are fluid in nature and have a mosaic organisation of patterns associated with protein coding regions, transcriptional regulatory regions, and repetitive sequence elements (Shapiro, 1992). Protein coding regions consist of domains that can be used many times and in a variety of combinations with other domains to produce mosaics. For example, in two-component bacterial transcription regulation systems, both sensory and regulatory proteins fall into families that share conserved domains (protein kinase, phosphorylation substrate, and

Table 1. Two views of the genome (adapted from Shapiro, 1991)

Older View	Contemporary View
Constant genome	Fluid genome
Rigid storage	Dynamic storage
Occasional copying errors and physico-chemical accidents	Monitoring, correcting, and change by dedicated biochemical complexes
Bag of individual, isolated genes	Multigenic systems
Information utilization in automatic/mechanical fashion	Integrated, coordinated, and complex information system
Mechanical-chemical	Information-rich ("smart")

DNA binding) but differ in other domains, according to specific regulatory tasks.

Magnasco and Thaler (1996) argue that certain heritable mechanisms could change the rate of evolutionary novelty. The genes for the enzymes involved in proofreading, mismatch correction, and DNA repair are themselves subject to intrinsic and extrinsic changes. Rather than being solely a stochastic and accidental process, genetic change is the result of adaptive feedback. The importance of networks of feedback information in the processing of hereditary information is a central feature underlying the idea of "knowledgeable" genetic systems (Holmquist & Filipsk, 1994; Thaler, 1994). At the level of ecosystems, hereditary information may not only flow across the generations of a species, but also between different species within a few generations. Richmond (1979) uses the metaphor of a cell or organism as a habitat for DNA to discuss the transfer of chromosomes, plasmids, and transposons within bacterial populations. Mazodier and Davies (1991) describe the inheritance of antibiotic resistance between distantly related bacteria in terms of bacteriophages acting like ecological agents with the niche of DNA-syringes.

Although individual-based thinking has been facilitated by increased computer power and accessibility of the techniques involved, a major reason for the shift is that the individual-based approach accounts much more fully for the intricacies and details considered key in determining how an ecological system functions. This approach can be used to make specific predictions about systems where data can be collected at a fine enough individual and spatial resolution to validate models (e.g., Keeling & Gilligan, 2000). The systems we shall describe in subsequent sections provide environments in which artificial (virtual) bacteria are simulation models for selected aspects of real-world microbiological agents. The computational agents allow individual and adaptable changes to be traced over time. It is possible to explore what happens in a system in which there are very large numbers of individuals interacting with each other and their environment over time and space. Energy and material are conserved in these systems, and information has an associated cost with regard to its production, translocation, and transduction. With regard to all models, we acknowledge that there is a trade-off between realism, precision, and generality. As such, any given architecture will emphasise some aspects of the real world and de-emphasise others. We have explored a number of different agent architectures, including a fine-grained model of a cell, in which individual molecules or functional agencies are represented as software objects, and a coarser-grained model that looks at the ecology, adaptability, and evolvability of bacterial behavioural strategies.

A SYSTEMS BIOLOGY FOR EVOLVING BACTERIA

The Virtual Cell project (Schaff et al., 1997; Schaff & Loew, 1999) makes use of user-defined protein reactions to simulate compartments at the nucleus and cellular level. GEPASI (Mendes, 1997) also models protein reactions, but from within an enclosed box environment. The BacSim project (Kreft et al., 1998) simulates individual cell growth at the population scale. Eos (Bonsma et al., 2000) is also based at the population scale but is intended as a framework for testing idealised ecologies, represented by evolutionary algorithms. The E-Cell project (Tomita et al., 1999) aims to use gene data directly in a mathematical model of transcription. All of these models can be good at modelling their own carefully specified niche. However, they suffer the same drawbacks — all rely on actual experimental data (which can be incomplete) and, more importantly, once input, those data are static. As a result of this data problem and the highly specific goal of the models, it became clear that the required model of evolution did not exist as an extension of these models.

A successful bacterial model must encompass the important qualities of bacterial evolution and bacterial life, but not be over-specified so as to constrain the results. Such a computational model must encompass some concept of genetics, a population, and an environment that can include spatial characteristics. Described in this section is the model known as COSMIC I that fulfils these criteria by carefully balancing biological and computational realities (Way, 2001) with an emphasis on open-endedness (Kampis, 1996) and individuality, while also operating within a parallel implementation. This chapter aims to show that a cell can be viewed as computational, and while using that computation, can evolve the connections that make up the computation to perform the chosen task of hill-climbing a substrate gradient.

The main aim of COSMIC I is evolutionary modelling based on biologically realistic organisms. This process includes interactions within cells involving the combined effects of enzymes and regulatory proteins acting on genes, which in turn act on those enzymes and other proteins, creating a huge number of both positive and negative feedback loops necessary for cellular control. All of the concepts implemented in this chapter use an individual-based modelling philosophy: the environment contains individual cells, each cell contains an individual genome, and each gene can lead to individual gene products, each with its own spatial and temporal parameters. This vast number of parameters (both static and those seeding the attribute values associated with each object) and possibilities adds another meaning to the name of the simulation: COmputing Systems of Microbial Interactions and Communications, or COSMIC.

Part of the central dogma of molecular biology is that the transcription process is carried out by an enzyme specific for the creation of mRNA species on a DNA template. This enzyme (RNA polymerase) can be thought of as a complex protein-based machine that performs the task of bringing together the constituent parts and making mRNA. To start transcription, an RNA polymerase molecule binds at a promoter site (a transcription starting point), and the RNA polymerase then transcribes as it moves along the genome, stopping at a terminator site. An interesting feature of this system is variation in gene expression. As enzymes and regulatory proteins may be used to control transcription in a region of DNA, they may be either transcription repressors or activators, and they may themselves be controlled by other regulatory proteins.

Operator regions form sites onto which repressor proteins (DNA-binding proteins) can attach. Once attached, the mRNAs encoded by that region are not transcribed, and hence the protein is not manufactured. In this case, the repressor essentially gets in the way of the RNA polymerase. If an inducer molecule is present, it can bind to the repressor molecule and nullify the repressor's effect. Repressing or activating proteins reflects both the genotypic and environmental state of a bacterial cell, creating rapid, complex processing.

A Model Incorporating the Genome and the Proteome

There is a direct mapping of these biological concepts into an object model that supports evolutionary processes. The key goal is not accuracy (which is clearly impossible in an evolving system that would need constant updating of parameters from real-world data — data that do not exist), but expressibility, which amounts to a framework for evolution. Although a simple model of a genome can be represented by a string of letters, there are layers of interpretation that can be placed on top of this representation. Looking at genome maps (e.g., *www.ecocyc.org/*) shows that the strings are divided into nonuniform lengths, each of these identifying some gene or other active string sequence. Sequences can be broadly categorised into those that encode an enzyme, and those that act as regulatory structures on which enzymes (or further nucleic acid sequences) act (Record et al., 1996; Collado-Vides et al., 1998). As a result, the genome in COSMIC I dynamically encodes eight broad sequence types derived from those present on the *lac* and *trp* operons and allows for multiple interaction types of single genes.

These eight sequence types are the DNA binding sites, promoters, operators, attenuators, and receptors (including FAPs — flagella activation protein sequences), and the protein agents' inducers, repressors, and RNA polymerase (here called sigma factors). Receptors provide the interface between the optional transcription cascade and the cell's external environment. Of these eight there are four fixed types: promoters, operators, attenuators, and a type that marks a gene as transcribable, but with an as-yet-unknown purpose. Gene-gene and gene-enzyme interactions then follow from pairing fixed types to these as-yet-unknown types that were originally defined to be only some kind of transcribable gene. These relationships are shown in Figure 1. Interaction pairing is based on the antimatch of the DNA-like encoded strings contained in each gene object and so responds to the changing genome. But it is also stable over time when the genome is stable.

COSMIC I realises the genome and gene products using the individual-based modelling approach, where each molecule is considered as an individual entity with individual attributes. This approach allows for the inclusion of spatial effects (RayChaudhuri et al., 2001) without imposing artificial spatial boundaries of fixed resolution and without modelling only mean quantities. Given a population of gene products in the cell, each has a chance of reacting with each other and on the genome according to Figure 1. The probability involved is based on the gene product's type as defined by the genome, its position, its half-life, and its age. Each reaction lasts for a variable time based on the type of gene pairing. The goal of this probability distribution is to be unbiased, giving each of these attributes a chance to have some effect without deliberately allowing any single attribute to dominate.

Figure 1. Enzyme type interactions

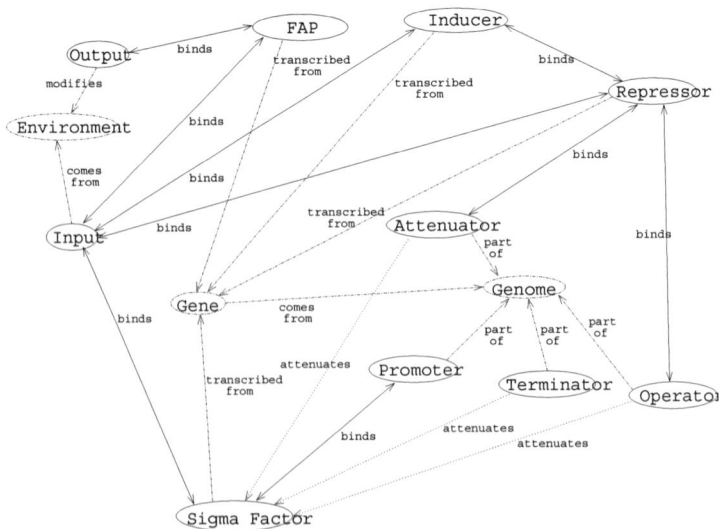

As an example, the repressor-operator interaction type consists of many instances that relate genes and gene products, the link based on the antimatch function of the gene sequence. Each relation of this set of reactions is then tested one at a time using the above parameters individual to that part of participating objects. This test assigns a probability of occurring, and a random number is then compared to this probability which decides whether the relation will act and so bring the gene product (repressor) and gene (operator) together. In biological reality, the repressor must be actively removed and so needs no further probability functions, but the relationship for other relations is based around this probability for both binding and unbinding reactions.

The case of promoter-sigma factor interaction is more involved as this is the source of gene products. Given a valid antimatch between promoter and gene, that gene becomes known as a sigma factor, in the same way as mentioned earlier. Once a successful probabilistic match has taken place, the sigma factor binds the promoter site but then unbinds. The role of the sigma factor is merely an instigator of transcription; an imaginary RNA polymerase then moves along the genome starting at that promoter site, transcribing genes as it moves. Compared to RNA polymerase, in the real cell sigma factor is in short supply and so there is no need for a literal RNA polymerase.

The fully specified COSMIC I model (Gregory et al., 2004a, 2004b) consists of a hierarchy of sets of objects. The most basic set type is the gene, which is part of a genome set type, which is part of a cell set type as part of the environment. The cell set also contains the enzyme set, which is populated by gene products of the genome. Each level contains additional parameters and attributes associated with that level. For instance, the spatial information that partly computes reaction probabilities is contained in each gene and gene product. A conceptual view of the components in an individual cell is shown in Figure 2.

When specified for all genes, these relations between genes and gene products give a picture of which gene products (enzymes) can react. Not all sequences are genes — control sequences are also considered in the antimatching process. This process creates interaction paths between control sequences and gene products, thus allowing these gene products to then be called sigma factors, repressors, and inducers. This forms a set of possible reactions between genes and gene products. Other sets of relations pair reactions that are occurring with certain elements of certain sets.

This system also requires some form of sensing of and responding to the environment. For this sense, COSMIC I considers receptors on the cell wall (input) and flagella motion (output) to be directly related to the transcription

Figure 2. Conceptual view of a cell

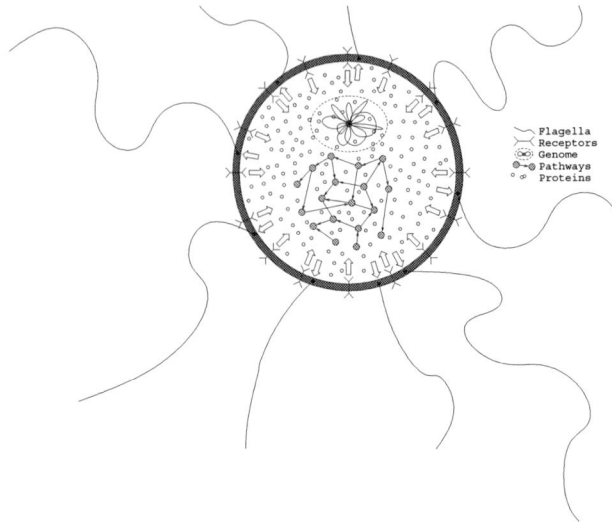

networks. The receptors represent imitation gene products that exist in fixed positions on the cell wall and have an infinite half-life. These gene products can bind to normal gene products using a set of relations also based on the antimatch function. The flagella response (output) follows in the same way. COSMIC I records in full detail the cascade of interactions between gene products and genes, which can then be converted into interaction networks.

For COSMIC I to function efficiently, the real biological process has been simplified to have no explicit molecule type on which enzymes react. A single type of generic pseudo protein is assumed to exist, and all reactions act on that protein. This protein can be bound by enzymes, as well as by the normal process of enzymes acting on each other. The resulting effect is that the enzymes and the single FAP type are not available to do anything else while bound and so allow other molecules the chance to bind.

The Environment

The largest scale in COSMIC I is the environment. An environment is necessary for simulating bacteria by providing the means by which fitness can be implicitly enforced. Fitness is defined here as a measure of genome convergence towards the ideal of following a food gradient. The population of cells must compete side by side for what exists in the common environment. To this end, all cells live in a glucose-rich environment that is depleted over time by the cells (Kreft et al., 1998). Better cells in better regions of the environment will grow faster and so multiply faster. The combination of input reward based on cell position, and cell position based on flagellum output, produces an indirect reward-based system that is the basis on which the simulated *E.coli* evolve. The changing nature of the environment ensures that there is no single ideal and that any better solutions may become inferior over time, thus creating a fitness function that adapts to the environment.

Figure 3 shows a snapshot of the environment in which white equates to a glucose level of 4.5 mg, and black equates to an absence of glucose. This area is 0.2 mm square. Black circles represent bacteria that have not moved through lack of connection with their flagella, and grey streaks show moving bacteria. Glucose use for each cell has been exaggerated to better motivate motility and so evolutionary change.

At this early stage in the simulation, the genomes still have little order, having grown for approximately nine generations, leaving little genetic mutation. As a result, the evolved behaviour is limited to uncontrolled movement. The most motile cells leave a lighter trail of glucose use and can be picked out visually as cell 0246 on the bottom left. Moving more slowly is cell 0248 on the bottom middle. Failed motility can be seen in cell 0230 on the bottom right. Black points without outline circles are cells that have since been removed through death, leaving only the environmental effects.

Figure 3. Environmental change after 166 minutes

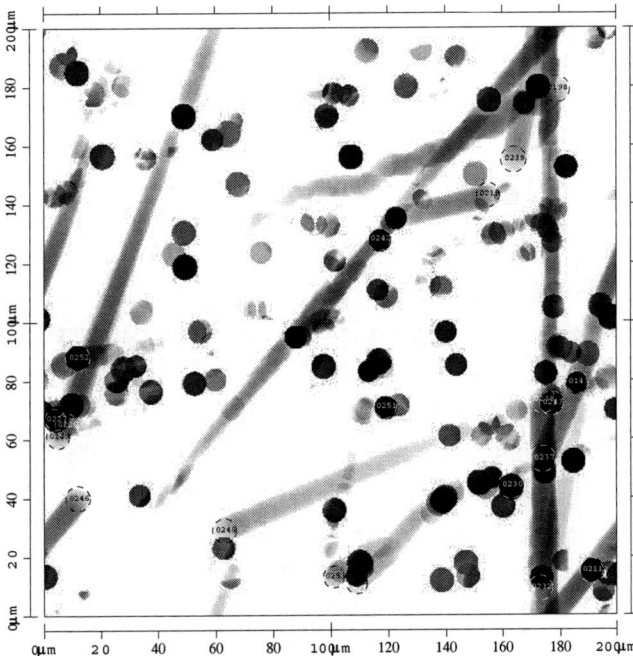

Evolvability is the goal of COSMIC I, not prediction of known behaviour. The generality needed for evolution requires the simulation set's genetically determined parameters (e.g., transcription rates or reaction rates) as either random or unbiased. Verification then becomes a moot point. In the real biological case, there is nothing to compare the system to, and some data are available on mutation rates, on genome sequences after mutation, and on phenotypes of mutants. As a result, validation of this system is based solely on verifying the correctness of the parts. It is known quantitatively that the cell growth equations are correct, and that the concepts of the active genome exist and have a qualitative and, in some cases, quantifiable effect. What is not known is how these parts fit together; the legacy of practical biology leaves COSMIC waiting for a distant future of biology. There is so much more detail available from COSMIC at every level of measurement than there could be from years of wet lab research. This then makes full validation near impossible as there is nothing to compare to when the measurements do not exist. What can be verified are the behaviour of the concepts and the cell growth rate equations. Related concepts, such as the interaction paths, must and do execute fairly, ensuring no program error gives an interaction path higher priority over another. In all other respects we must be careful to avoid simulation artifacts.

To summarize, a model consists of a dynamically sized population of cells, each containing a genome and a proteome. The genome is a set of ordered genes, ordering being necessary for the emulation of transcription. The proteome is a set of gene products that act as enzymes and genome-reading machinery. Together the genome and the proteome create a self-sustaining machine. Without mutation, the genome is static. The proteome is always dynamic, populated with the transcription products of the genome, but reduced in population by cell division and enzyme half-life. The two normally distinct areas are unified with many simplifications that must also allow for the continuous reinterpretation of the genome and continuous updating of the proteome. This reinterpretation of the relationships encoded by the genome is essential as it drives adaptive behaviour, and it is that adaptive behaviour that we call evolution.

COSMIC I executes using a parallel architecture (Gregory et al., 2004a) under the client-server paradigm, as this enables a reasonable computation time for a given evolutionary time frame. The client-server paradigm is used at a coarse-grained time step, each step being the opportunity to synchronise past events between server (the environment) and clients (cells). The fine-grained time is the intracellular interactions between the genome and the proteome. The synchronisation here is in biological time rather than wall-clock time and so ensures that all cells execute for as long as they require to test for and execute their internal interactions.

Searching for Evolution

There are clearly two distinct scales to the COSMIC I system: the state of the environment and the state of each cell. Some of the resultant data have already been presented in this chapter, but the search is really for signs of evolution.

Simulations have shown that the slowly evolving population makes use of the dynamic genome size, duplication, and deletion events, creating a much more diverse genome size. For each cell, genome size is initialised to within 70 and 120 genes (total of both control genes and transcribable genes) with a mean of 96 genes. In over 2,200 minutes of simulation time, the variance increases so that the mean is 800 genes, with a minimum of 70 genes and a maximum of 20,000 genes.

In conjunction with the genome size, the variance in gene products per cell also increases over the course of the simulation. The high variance suggests that gene products are controlled by optional transcription. This transcription itself is constrained during cell division, where a high sensitivity to the division of gene products can massively change the production of further gene products. A massive reduction normally ends with the termination of the cell as it is classified as no longer viable. For new cells with no parent, enzymes are initialised to have five enzymes per transcribable gene. At the end of the simulation, the mean enzyme concentration over all cells was 10,000, the minimum was 0 (which would have quickly led to the cell being terminated by the implementation), and the maximum was 100,000 enzymes, presumably the result of a futile cycle. These figures suggest there is pressure for a larger genome and more active cytoplasm. The main evolutionary pressure is to encourage the cells to be more robust when subjected to cell division.

In summary, the COSMIC I model presented here is a simulation of a bacterial population that departs from the norm by being extremely detailed, and at the same time, specifically targeted at evolution. The main focus here is on evolutionary change rather than reproduction of known individual phenomena. Evolution is implemented in an object-based model of optional transcription, including a genome and a proteome, with a population of cells that are genetically independent but share a common environment. The object-based approach has enabled the use of individual modelling and the capacity to track the contributions made by individual cells to the evolving processes.

The COSMIC I system represents one of the most all-inclusive simulation systems of bacterial evolution, incorporating biological processes that normally exist in separate models (i.e., transcription cascades, ecology of multiple organisms, and evolution of these organisms). Included in this system are interactions within and between the genome and proteome from the bottom up, with no predefined transcription paths. Each participating object is not only encoded as a software object, but it can also be directly mapped to realistic proteins and genes, though that requires some degree of simplification regarding the relevant biophysical processes as COSMIC is not a chemical simulation. These interactions are less complex than their biological analogues in terms of types, yet these collections of interactions enable a collective or systemic observation as well as the individual, object-based observation. Above the viewpoint of the cell is the collection of cells in the environment, which gives the most high-level view. Evolvability, not prediction, is the end goal, and these different viewpoints give COSMIC I the scope to record and find the root causes, with the possibility of then inferring the same causes in the real bacterial analogue.

The results gathered from COSMIC I so far show that evolution could well be taking place despite the short simulation time and the vast effective search space. The computational effort is significant and stands to limit the evolution of COSMIC I itself. It has been tested on a Linux farm and a 12-node dual processor Athlon XP 2000+ Beowulf Cluster, both running under PVM. On the latter, it produced 3 GB per day of simulated data. A run of a week simulated approximately two days of bacterial life and tested up to 490 simultaneous cells. By far the biggest hurdle is the sheer amount of data generated. This quantity of data leads to further stages of COSMIC I development and tools to analyse the output data in a meaningful and concise way and to move COSMIC I to a grid environment. These tools could well be useful in the context of more traditional systems biology approaches.

A RULE-BASED APPROACH TO BACTERIAL EVOLVABILITY

The COSMIC I system operates at the detailed level of networks of gene-protein interactions. It bootstraps random genomes and then tests them in a nutrient-depleting environment that acts as a fitness function. COSMIC I is computationally very demanding, and as we continue to develop it we are also conscious that to model the complexities of biological evolution with large numbers of bacteria and longer run times, it will be necessary to explore greater computing power and simplify or abstract behaviours of the individual agents. In this section, we describe one way this problem has been addressed, using an evolvable, rule-based bacterial modelling method.

This approach to bacterial modelling employs a rule-based architecture to model some of the complex environment-organism interactions that can occur in natural bacterial ecologies. Essentially, bacteria are treated as information processing units that map external "messages" coming from the environment to appropriate "actions" to be taken. These message-action mappings are not static, but are continuously evolving by means of mutations, thus enabling the artificial organisms to adapt to changing environments, maximise their lifespan, and improve their reproductive success. The derived model is known as RUBAM, which stands for RUle-based BActerial Modelling. The purpose of the RUBAM system is not to be an accurate model of any specific real-world ecologies or bacteria, but rather to capture prominent features of their behaviours, so that it can exhibit adaptive and evolvable patterns. Other models with similar features have been developed in the past, most notably Echo (Hraber et al., 1997; Holland, 1992) and Herby (Devine & Paton, 1997).

Structure of the RUBAM System

Even the simplest real-world bacterial ecologies are extremely complex in the context of computational modelling. This makes it practically impossible to develop a bacterial ecology model that is both accurate and realistic by conventional computer systems. In designing the RUBAM system, a careful balance had to be maintained between model complexity and computational demands. The model was kept simple enough to be computationally realistic, while being complex enough to be able to exhibit adaptive and evolvable behaviour.

The RUBAM system consists of a number of fundamental elements, which work together to construct an artificial ecology. These include an artificial environment, a collection of artificial organisms, a set of environment-organism interaction mechanisms, and a set of evolutionary operators.

The Artificial Environment

The artificial environment can be thought of as a "container" that contains every other element in the system. It is represented by a n-dimensional grid, such as the one shown in Figure 4 (for a two-dimensional space). The grid contains a number of uniformly spaced grid points, known as environmental sites, in which the artificial organisms are placed to survive, interact, multiply, and evolve.

*Figure 4. Environmental grid in two dimensions, and environmental site
with three resource types*

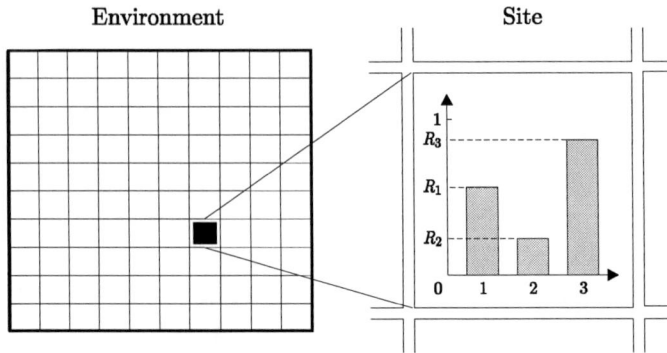

The environment can contain any number of artificial resources in various concentrations. Resources are an important element of RUBAM, as they provide the energy that is required to sustain life. Different types of resources can exist in the environment, and each type can have a different effect on an organism. The system is designed to conserve matter and energy, which are conflated for the purpose of tractability. Most importantly, the resources (in certain combinations of concentrations) can trigger different types of behaviour in the organisms. Depending on the degree of fitness of the organism to its local environment, this behaviour can be destructive to the organism, or it can generate a desired effect, in which case the organism gradually accumulates enough energy to multiply and generate copies of itself in the environment. In this way, those organisms that are able to better exploit the resources have a higher chance of propagating through to successive generations and of generating more copies of their genetic material in the population. Figure 4 shows an example of an environmental site containing three resource types in different concentrations: $R_1, R_2,$ and R_3.

The Artificial Organisms

The artificial organisms are at the heart of RUBAM and collectively contain all knowledge gained through evolution. They are modelled using the concept of a learning classifier system (LCS). An LCS is a rule-based system (also known as an agent or, in this case, virtual bacterium), that is able to receive messages coming from external sources (the environment or other organisms present in its neighbourhood), process them in some way, and generate appropriate actions that can affect the organisms themselves, as well as the outside world. Although LCSs were not specifically designed with evolutionary modelling in mind, they fit well within the scope of rule-based bacterial modelling. More details on LCSs and their applications can be found in Holland et al. (2000) and the references therein. The "genetic material" of each organism consists of a collection of rules, which map its "inputs" (known as detectors) to its "outputs" (known as effectors). This set of rules determines the behaviour of the organism when subjected to different types of stimuli. The objective is to modify the rules in such a way that a desired behaviour is obtained. In the context of evolution, "desired behaviour" is understood as the behaviour that maximises the lifespan and reproductive success of an individual. An illustration of the most important elements of an artificial organism in RUBAM is shown in Figure 5, and a brief description of each element is given in the following text.

The detectors provide a means for the organism to sense the state of the environment in its neighbourhood. This information is transferred to the organism by means of messages. The messages come primarily from resources that exist in the site in which the organism currently resides. Furthermore, the organism is designed to be aware of the gradients of the resource surfaces in its neighbourhood and of its current energy level. By "awareness" we mean that the organism's actions are not merely functions of the resource concentrations, but also of their gradients and the organism's energy level. These factors have the potential of enabling the organism to adopt different strategies as its energy level rises or falls, and as the resource landscapes become steeper of flatter.

The messages received by the detectors are sent to a rule-based classifier system, where they are matched to the antecedents of the rules in a rule base. This rule base constitutes the "genetic material" of the organism and directly affects its behaviour in response to those messages. There are many ways to perform the rule matching. In the RUBAM system, a fuzzy classifier system was used (Zadeh, 1965; Mamdani & Assilian, 1975). The rules in the rule base have the following standard format:

IF <antecedents> THEN <consequents>

The antecedents and consequents of a given rule contain a number of parameters that define the input and output fuzzy sets that correspond to that rule. The messages are formed by the resource concentrations, their gradients along the n dimensions of the environmental grid, and the organism's energy level. All of these crisp values form a message vector, which is passed to the fuzzy classifier system. This vector is then fuzzified with respect to every rule,

Figure 5. Artificial organism in the RUBAM system, with its most important elements

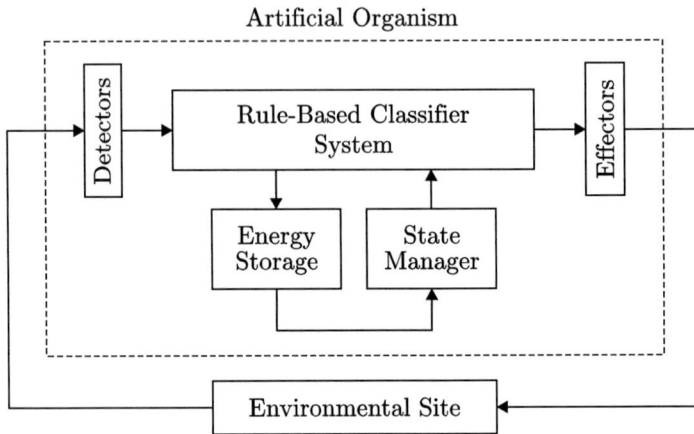

and the corresponding output fuzzy sets are determined. The defuzzification method proposed by Mamdani and Assilian (1975) is then used to defuzzify the output fuzzy sets and infer a set of crisp values that are passed to the effectors and turned into actions. Even if there are conflicting rules, the method in Mamdani and Assilian (1975) ensures that a set of crisp output values will always exist, but this method can introduce redundancy in the rule base (i.e., two or more conflicting rules that balance each other out). This redundancy can be minimised by penalising organisms with a large number of rules.

Each organism contains an energy reservoir, which accumulates energy from the resources and then uses it to sustain life by performing appropriate tasks. The level of energy stored inside an organism is bound in the interval [0,1] and controls the ability of that organism to receive messages and the way in which it responds to those messages. This is the purpose of the state manager, which monitors the organism's energy level and brings the organism to a number of different states. For example, if the energy level is too low, the organism is unable to detect messages or perform actions and is said to be in the inactive state. Conversely, if the energy level is sufficiently high, the organism is able to initiate the division process and may eventually divide into two copies of itself. One of the offspring remains unchanged and identical to its parent, while the other is subjected to a set of evolutionary operators that modify its rule base in some way, thus enabling it to evolve. A commonly used evolutionary operator is the mutation operator, which randomly alters the symbols used to describe the rules in the classifier system's rule base with a small probability.

The effectors receive commands from the consequents of the matched rules in the rule base and perform appropriate actions. Typical actions include the following:

- Movement to a different neighbouring site in the grid.
- Change of organism's energy level (positive or negative).
- Resource generation.

Certain actions can directly affect the environment (i.e., resource generation), while others can also modify the organism itself (i.e., movement, change of energy level). Resource generation is an important action in that it enables the organisms to "communicate" with each other. This communication is possible because an organism is potentially capable of receiving messages generated from resources that another organism has produced. Evidence of coordinated behaviour among certain types of bacteria has been observed in real ecological systems and is known as "quorum sensing". This mechanism enables certain bacteria to regulate their population densities by detecting the presence of other bacteria in their neighbourhoods. The RUBAM system has the potential to create behavioural patterns similar to quorum sensing.

Evolution takes place by means of a set of evolutionary operators. These operators are stochastic in nature and alter the behaviour of the organism in some way by modifying its rules. There are several schemes that could be employed, with mutation (in various forms) being the fundamental evolutionary operator. The behaviour of an organism can also be altered by modifying the "importance" of each of the rules in the corresponding LCS, based on their observed performance. A popular mechanism for modifying these rules is the bucket-brigade algorithm (Holland, 1985). Other reinforcement learning algorithms can also be employed. An overview of reinforcement learning and its applications can be found in Kaelbling et al. (1996) and the references therein.

Another important element in RUBAM is the choice of logic that is used to map the messages received by the detectors to actions taken by the effectors. The proposed system is capable of using both traditional (Aristotelian) logic as well as fuzzy logic (Zadeh, 1965). In this work, fuzzy logic was employed because it generally results in rule bases that are easier to comprehend and express in natural language (Zadeh, 1965). The inference method used to drive the effectors of the organisms was the one proposed by Mamdani and Assilian (1975).

The Organism's Different States

Depending on its energy level, an organism may be in one of four different states, which can affect the way in which it responds to messages and performs actions. The range of possible energy levels (between 0 and 1) is divided into four regions, each one corresponding to one of the following four states:

- Inactive.
- Active.
- Division preparation.
- Division.

When the organism is in the inactive state, it cannot normally receive messages from the environment and cannot normally perform actions. The organism enters this state when its energy level has dropped below an inactivity threshold, and is not sufficient for its normal functioning. The organism is, thus, inactive. In most cases, the organism cannot leave the inactive state once it reaches it, which means that the organism effectively "dies". However, it is possible to define conditions under which an organism in the inactive state can "wake up" and become active again.

The organism is expected to spend most of its lifetime in the active state. In this state, the organism has sufficient energy to function normally (i.e., it is able to respond to messages and perform actions).

Once the energy level of an organism exceeds a threshold, the organism enters the division preparation state. In this state, the division process is initiated, and the organism prepares to divide. This state usually functions similarly to the active state, although in some cases it may inhibit the organism from performing certain actions.

The division state is a transient (unstable) state. Once the energy level of an organism exceeds an upper limit, the organism is able to successfully complete the division process and is split into *two copies of itself*. After the division process, each of these copies contains approximately half the energy of the "parent" organism. The two copies are, therefore, immediately brought back to one of the earlier states, most likely the active state. This means that an organism usually stays in the division state for only one time step.

The Mutation Operator

As mentioned earlier, once an organism manages to accumulate enough energy to reach the division state, it immediately divides into two other organisms, which are identical copies of their "parent". To enable the organisms

Figure 6. The division process and mutation operator

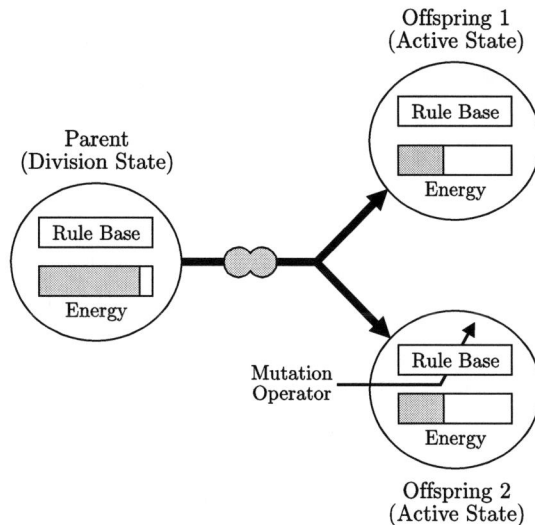

to evolve, a small amount of mutation is applied to one of the two "offspring" so that the original genetic material is preserved, and at the same time, new genetic material is introduced in the population. The mutation operator is similar to that of standard genetic algorithms. Each of the symbols that constitute the rule base of the organism to be mutated is altered with a given mutation probability. The mutation probability is usually kept very small, and is often chosen as a function of the rule base size. Figure 6 illustrates the division process and the application of the mutation operator to one of the offspring.

Simulation Results

This section presents a typical simulation run of the RUBAM system, on a two-dimensional grid (i.e., a plane) consisting of 100 x 100 sites, with an initial population size of 50 bacteria. As expected, the system is very sensitive to initial conditions and several attempts had to be made until the initial population contained organisms with sufficient fitness to initiate the evolutionary process. To maximise the adaptation of the organisms, several simulation runs were conducted, with each successive run initialised with the best performing organisms of the previous run. We define "best" as those organisms with the longest lifespans and the highest number of offspring.

Figure 7 shows a snapshot of the final simulation run of this experiment. Dark-to-light areas correspond to low-to-high concentrations of resources.

Organisms with moderate energy levels (in the active state) are marked with small circles, those with high energy levels (in the division preparation state) are marked with large circles, and those in the inactive state are marked with crossed circles. Three different resource types were used in this simulation, with the first one occupying the top left region of the environment (wide bell-shaped surface), and the second one occupying the top right and bottom left regions (narrow bell-shaped surfaces). The third type was used as a signalling resource (i.e., generated by the organisms) and is not shown in Figure 7. During this simulation, the organisms evolved ways to "climb" the hills formed by the resources to reach nutrient-rich areas in the environment, which in turn enabled them to grow and reproduce. At the end of the final generation (generation 99), the population size increased from 50 organisms to 840 organisms. The paths followed by some of them can clearly be observed in Figure 7, leading to areas of high resource concentrations. The lifespans of all 840 organisms during the course of the final simulation run can be observed in Figure 8. As demonstrated in Figure 8, some organisms die (become inactive) quickly, while others are able to stay in the active state for longer and eventually reach the division state and reproduce.

The top of Figure 9 shows the variation of the population sizes for the total, active, and ready-to-divide organisms during the course of the simulation. The

Figure 7. Snapshot of the environment taken during a simulation run of the RUBAM system

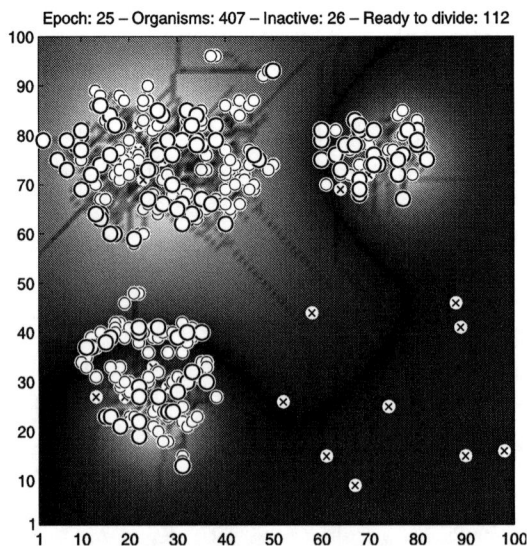

Epoch: 25 – Organisms: 407 – Inactive: 26 – Ready to divide: 112

Figure 8. Total lifespans of each organism in the simulation run of the RUBAM system

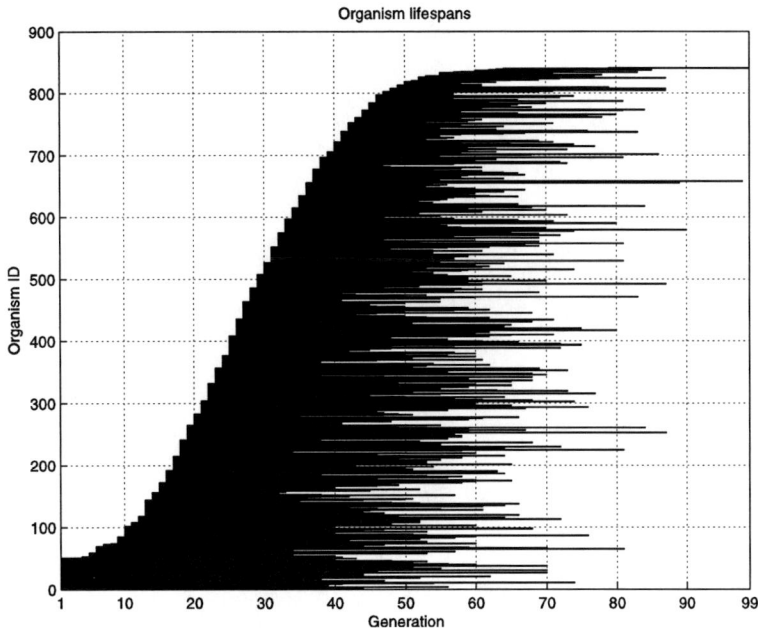

Organism lifespans

ability of the organisms to exploit the resources is better observed in the bottom of Figure 9, which shows the variations in the total concentrations of all three resources during the course of the simulation. As the figure illustrates, the organisms are able to exploit resource types 1 and 2 (their concentrations are steadily decreasing from generation to generation) until they are no longer sufficient to provide the organisms with enough energy to stay in the active state. The organisms then become inactive and the simulation terminates at generation 99. An interesting observation is that resource type 3 (the signalling resource that models quorum sensing) increases steadily, which indicates that the organisms generate signals for other organisms to receive. This suggests that the organisms may have evolved ways of communicating with each other and thus coordinating their behaviour in some way.

Upon examination of the rule bases of the fittest organisms in the population after the final run, the evolved behaviour closely resembles that of a typical hill-climbing optimisation algorithm, namely the movement towards the direction of ascending gradients of the resource surfaces. This result, although relatively simple, demonstrates the ability of RUBAM to evolve behavioural patterns that are applicable to real-world problems, such as function optimisation

Figure 9. Population sizes and normalised resource concentrations in the simulation run of the RUBAM system

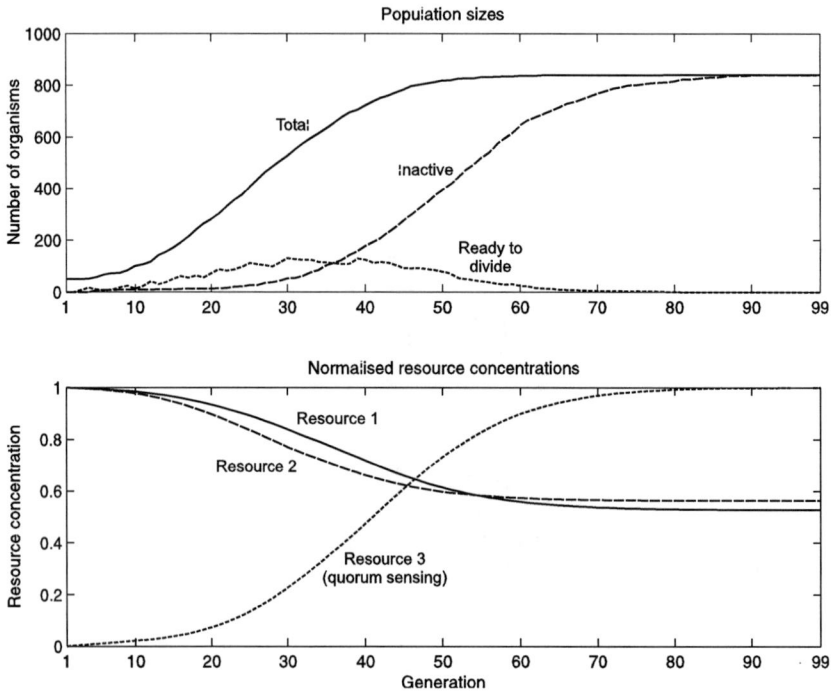

problems. Other RUBAM simulation runs have generated strategies that can help locate multiple optimal solutions in multimodal search spaces. Furthermore, RUBAM allows the use of environments of higher dimensions (i.e., $n > 2$), thus enabling the investigation of abstract environments that may not necessarily exist in the natural world but can be used to solve particular problems.

It should be stressed at this point that RUBAM is neither a genetic algorithm (GA) approach (Goldberg, 1989), nor a genetic programming (GP) approach (Koza, 1992). There is no external objective function being optimised as such, but rather, a fitness measure that is *internal* to the individual organism. In other words, the fitness of an individual can only be assessed by subjecting it to an environment and letting it interact with it for some time (generally until its energy reserves are exhausted). In contrast to that, in standard GA/GP approaches, the fitness of an individual can immediately be computed by means of a known and static objective function. Initial results have shown that RUBAM can generate solutions that are fundamentally (and favourably) different from traditional GA/GP approaches. A disadvantage of the proposed approach, as with most evolutionary approaches, is that it is computationally intensive. The possibility of implementing RUBAM in multiprocessor computer environments is currently being investigated. The following section highlights one aspect of this move towards greater computational power with regard to developing more complex (and challenging) evolutionary models.

COMPUTATIONAL GRID SYSTEMS FOR SUPPORTING SIMULATION ENVIRONMENTS

As the number or complexity of agents increases in a model, a natural consequence is that the processing time can become a limiting factor in obtaining useful results, and ways of reducing the load on a single processor will inevitably need to be found. The scope for improving the efficiency of each agent process (for instance, by improving the mathematical algorithms used as noted in Haefner & Dugaw, 2000) is usually limited, and simplification strategies such as aggregating population groups of individual agents may defeat the whole object of using "individual" agents in the first place, unless special precautions are taken (see, for instance, Scheffer et al., 1995). In these circumstances a "distributed computing" strategy, where the computing load is spread amongst a number of independent but interconnected processors, may prove the only viable alternative.

In recent years, so-called "grid computing" has become an increasingly popular variation on the distributed computing theme. At the heart of a "grid" is the ability to link multiple, diverse (in size, architecture, and operating environment) computing engines or storage servers together to allow their resources to be shared in a seamless and transparent manner. Over time, the actual resources simultaneously available within the grid will change dynamically as individual members enter and leave the federation. This requirement for a fluid, ill-defined, changing network of resources with a uniform unchanging interface is what differentiates the grid from traditional distributed or parallel computing. Although the idea of using underutilised CPU cycles on interconnected computers has been around since the early 1970s, Foster and Kesselman are generally recognised as having formally defined the grid in the late 1990s, most fully in their book *The Grid — Blueprint for a New Computing Infrastructure* (1999), and more recently with Tuecke (Foster et al., 2001).

In principle, individual-based models (IbMs) lend themselves to a distributed computing approach, as their natural design uses many similar objects and processes, which can clearly be shared between separate processors. Implementation is not without its problems though. For instance, how should processes be distributed? How do we synchronise model time steps on different hosts? What form should interagent communication take? What happens if a processor drops out of the grid?

An obvious way to parallelise an IbM is to distribute the individual subdivisions of the "environment" or agents between the networked computer hosts and make each host in the distributed system responsible for a fixed portion of these. Each computer host can be made responsible for a fixed set of individuals or agents, without regard for their position in space, or each computer can be made responsible for a given, contiguous spatial area of the simulated environment. In the former methodology, each agent needs to know about all of its neighbouring agents, which implies that each individual must have access to a complete map of all other individuals. In the latter approach, each computer is responsible for a given area of simulated environment and therefore many agents potentially only need to communicate with their immediate neighbours co-located on a single computer.

A good example of this second methodology is exemplified by the SIMPDEL model (Abbott et al., 1997). This model tracks the growth, movement, feeding, and breeding of individual deer and panther across South Florida, and runs on a 32-node Thinking Machines Corporation CM-5. An updated version of this simulation (Mellot et al., 1999) exhibited a speed increase of 3.83 on a network of four Sun Ultra 2 machines on an ATM network using MPI protocols.

Most parallel IbM models are currently implemented using "traditional" remote procedure call (RPC) processes, most notably by using such technologies as CORBA, EJB, and COM/DCOM, and often using MPI or PVM protocols. However, relatively recently, "space-based" paradigms (for an example, see Nugala et al., 1998) have appeared. In these paradigms, tasks or data are distributed via a virtual whiteboard or "tuple space". Client hosts or processes can take a task, perform it, and replace the completed task, with associated data, back into the space for subsequent collection and analysis. This technology makes use of a tuple space (or simply "space") that is effectively a virtual, distributed, shared memory. Probably the earliest implementation of this idea originated at Yale University (Gelernter, 1992) when the "Linda" programming language was defined. Linda comprised a small set of operations in combination with a globally accessible persistent store called the tuple space.

The essential operations used in space-based programming are extremely simple. In fact, only four methods are necessary: write, read, take, and notify. *Write* and *read* fetch and store objects in the space, *take* removes an object, and *notify* informs the system of a change in the space. This system provides a far simpler API for distributed computing than the more usual RPC model in which objects must synchronise their processes and pass information via explicit procedure calls. Increasing scale leads to very rapid increases in the number and complexity of "communication" paths required, thus producing highly complex systems that require careful planning and execution. In space-based implementations, by contrast, it is the "space" infrastructure that effectively synchronises all calls into it. In effect, it becomes a "bulletin board", providing a means of communication in which the participants:

- Are entirely independent of each other.
- Needn't share the same process, processor, or machine — they just need a common network.
- Need to know nothing about each other.

Space-based solutions are therefore easy to implement. There is a penalty to be paid in the form of a relatively high communications "overhead", particularly as the size of the space increases. However, these effects can be mitigated to a degree by the use of multiple spaces (Zorman et al., 2002), and by ensuring communication transfers are kept to a minimum.

The use of grid computing technologies not only allows for the practical acceleration of existing IbMs, but also allows for the development of more complex model representations. Some current work at Liverpool University (Palmer et al., 2004) is aimed at developing an IbM model that treats the environmental part of the IbM as a series of interconnected nodes, each capable of running on separate computers. The environment model allows local spatial properties (such as diffusion coefficients) to be easily defined and varied. In its present form, simplified "microbe" and source/sink agents have been implemented (which can themselves similarly run on independent computers), which interact with the environment model by way of "field" values interpolated from neighbouring environment node attributes. The source/sink agents allow environmental characteristics to be varied continuously with time, and later developments will allow modelling of mass transfer, or "fluid flow".

This topology, using Sun Microsystems's Jini and JavaSpaces Distributed Computing infrastructure (Halter, 2002), has certain advantages over the more traditional system, where the environment is modelled as a series of regular orthogonal cells arranged in a chequerboard manner. It allows considerably more flexibility in "describing" the environmental space, in that the arrangement of nodes is not prescribed, with the possibility that areas of complexity could have more nodes than areas where spatial characteristics are mainly homogenous and static. Other work (Rauch, 2003) has shown that a reasonably regular arrangement of such nodes can effectively model physical transfer phenomena with only local synchronisation (although this probably requires homogeneous processor speed and interprocessor communication connections). This implementation of an IbM allows for building arbitrarily complex environmental models and the integration of other computationally intensive agents, such as separate microbe simulation models.

CONCLUDING REMARKS

This chapter has discussed individual-based modelling (IbM) in highly computer-intensive projects concerned with modelling evolutionary and ecological processes in artificial bacteria. The IbM approach has been presented as a way of carrying out virtual experiments that are not possible in the laboratory. Our approach provides a biologically plausible method for exploring evolutionary "strategies" and how genomes might evolve. This is achieved without imposing anthropocentric constraints, but by using biologically relevant input. We are not seeking to best guess by entering all genetic inputs at the beginning (that is, developing a monolithic agent construction). In this virtual world we can retain, in computer memory, the virtual genome of any single ancestral cell for reference purposes and for reseeding further simulations. In the real world, with a single-celled organism, this transient information cannot be retained because the two daughter cells are produced at the "expense" of the parent. Moreover, an advantage of our system is that rates of evolutionary change can be accelerated *in silico* beyond the maximum generation times possible with even the fastest-growing bacteria.

ACKNOWLEDGMENTS

Support for parts of the work reported here has come from U.K. EPSRC (GR/R16174/01) together with U.K. DTI (e-Science Support / ESNW) to grid-enable one simulation system (THBB/008/00134C). Partners in the latter project are the MASA Group, IBM Life Sciences, and Esteem Computing.

REFERENCES

Abbott, C.A., Berry, M.W., Comiskey, E.J., Gross, L.J., & Luh, H.-K. (1997). Parallel individual-based modeling of Everglades deer ecology. *IEEE Computational Science and Engineering*, 4, 60-72.

Avise, J.C. (2001). Evolving genomic metaphors: A new look at the language of DNA. *Science*, 294, 86-87.

Bonsma, E., Shackleton, M., & Shipman, R. (2000). Eos—An evolutionary and ecosystem research platform. *BT Technology Journal*, 18, 24-31.

Bray, D. (1995). Protein molecules as computational elements in living cells. *Nature*, 376, 307-312.

Collado-Vides, J., Gutierrez-Rios, R.M., & Bel-Enguix, G. (1998). Networks of transcriptional regulation encoded in a grammatical model. *BioSystems*, 47, 103-118.

Conrad, M. (1992). The seed germination model of enzyme catalysis. *BioSystems*, 27, 223-233.

DeAngelis, D.L., & Gross, L.J. (1992). *Individual-based models and approaches in ecology: Populations, communities and ecosystems*. London: Chapman & Hall.

Devine, P., & Paton, R.C. (1997). Biologically-inspired computational ecologies: A case study. In Corne, D. & Shapiro, J.L. (Eds.), *Lecture Notes in Computer Science*, 1305, Berlin: Springer, 11-30.

Devine, P., Kendall, G., & Paton, R. (1996). When Herby met Elvis — Experiments with genetics-based learning systems. In Rayward-Smith, V.J., Reeves, C.R., Osman, I.H., & Smith, G.D. (Eds.), *Modern heuristic search methods*. Hoboken, NJ: John Wiley, 273-290.

Fisher, M.J., Paton, R.C., & Matsuno, K. (1999). Intracellular signalling proteins as "smart" agents in parallel distributed processes. *BioSystems*, 50, 159-171.

Foster, I. & Kesselman, C. (1999). *The grid: Blueprint of a new computing infrastructure*. San Francisco, CA: Morgan Kaufmann.

Foster, I., Kesselman, C., & Tuecke, S. (2001). The anatomy of the grid. *International Journal of High Performance Computer Applications*, 15, 200-222.

Gelernter, D. (1992). *Mirrorworlds*. Oxford: Oxford University Press.

Ginot, V., Le Page, C., & Souissi, S. (2002). A multi-agents architecture to enhance end-user individual-based modeling. *Ecological Modelling*, 157, 23-41.

Ginovart, M., Lopez, D., & Valls, J. (2002a). INDISIM, an individual based discrete simulation model to study bacterial cultures. *Journal of Theoretical Biology*, 214, 305-319.

Ginovart, M., Lopez, D., Valls, J., & Silbert, M. (2002b). Simulation modelling of bacterial growth in yoghurt. *International Journal of Food Microbiology*, 73, 415-425.

Goldberg, D.E. (1989). *Genetic algorithms in search, optimization, and machine learning.* Boston: Addison-Wesley.

Gregory, R., Paton, R.C., Saunders, J.R., & Wu, Q.H. (2004a). Parallelising a model of bacterial interaction and evolution. *Proceedings of the Fifth International Workshop on Information Processing in Cells and Tissues*, September 2003. To appear in *BioSystems*.

Gregory, R., Paton, R.C., Saunders, J.R., & Wu, Q.H. (2004b). COSMIC: Computing systems of microbial interactions and communications, 1 — Towards a modelling framework. In preparation.

Haefner, J.W. & Dugaw, C.J. (2000). Individual-based models solved using fast Fourier transforms. *Ecological Modelling*, 125, 159-172.

Halter, S.L. (2002). *JavaSpaces Example by Example.* Indianapolis, Indiana: Prentice Hall.

Holland, J.H. (1985). Properties of the bucket brigade algorithm. In Grefenstette, J.J, ed. *Proceedings of the 1st International Conference on Genetic Algorithms and Their Applications* (ICGA '85), Lawrence Erlbaum Associates, 1-7.

Holland, J.H. (1992). *Adaptation in natural and artificial systems: An introductory analysis with applications to biology, control, and artificial intelligence.* Cambridge, MA: MIT Press.

Holland, J.H., Booker, L.B., Colombetti, M., Dorigo, M., Goldberg, D.E., Forrest, S., Riolo, R.L., Smith, R.E., Lanzi, P.L., Stolzmann, W., & Wilson, S.W. (2000). What is a learning classifier system? In Lanzi, P.L., Stolzmann, W., & Wilson, S.W. (Eds.), *Learning classifier systems: From foundations to applications.* Berlin: Springer, 3-32.

Holmquist, G.P. & Filipski, J. (1994). Organization of mutations along the genome: A prime determinant of genome evolution. *TREE*, 9, 65-68.

Hraber, P.T., Jones, T., & Forrest, S. (1997). The ecology of Echo. *Artificial Life*, 3, 165-90.

Kaelbling, L.P., Littman, M.L., & Moore, A.W. (1996). Reinforcement learning: A survey. *Journal of Artificial Intelligence Research*, 4, 237-285.

Kampis, G. (1996). Self-modifying systems: A model for the constructive origin of information. *BioSystems*, 38, 119-125.

Keeling, M.J. & Gilligan, C.A. (2000.) Metapopulation dynamics of bubonic plague. *Nature*, 407, 903-906.

Keller, E.F. (1995). *Refiguring life: Metaphors of twentieth century biology.* New York: Columbia University Press.

Kendall, G. (1998). *An ecological framework for artificial vision.* Ph.D. Thesis, Liverpool University, United Kingdom.

Konopka, A.K. (2002). Grand metaphors of biology in the genome era. *Computers & Chemistry*, 26 (5), 397-401.

Koza, J.R. (1992). *Genetic programming: On the programming of computers by means of natural selection.* Cambridge, MA: MIT Press.

Kreft, J.-U., Booth, G., & Wimpenny, J.W.T. (1998). BacSim, a simulator for individual-based modelling of bacterial colony growth. *Microbiology*, 144, 3275-3287.

Kreft, J.-U., Picioreanu, C., & Wimpenny, J.W.T. (2001). Individual-based modelling of biofilms. *Microbiology*, 147, 2897-2912.

Levins, S. (1984). The strategy of model building in population biology. In Sober, E. (Ed.), *Conceptual issues in evolutionary biology.* Cambridge, MA: Harvard University Press, 18-27.

Magnasco, M.O. & Thaler, D.S. (1996). Changing the pace of evolution. *Physics Letters A*, 221, 287-292.

Mamdani, E.H. & Assilian, S. (1975). An experiment in linguistic synthesis with a fuzzy logic controller. *International Journal of Man-Machine Studies*, 7, 1-13.

Mazodier, P. & Davies, J. (1991). Gene transfer between distantly related bacteria. *Annual Review of Genetics*, 25, 147-171.

Mellot, L.E., Berry, M.W., Comiskey, E.J., & Gross, L.J. (1999). The design and implementation of an individual-based predator-prey model for a distributed computing environment. *Simulation Practice and Theory*, 7, 47-70.

Mendes, P. (1997). GEPASI: A software package for modeling the dynamics, steady states and control of biochemical and other systems. *Trends in Biochemical Science*, 22, 361-363.

Nugala, V., Allan, S.J., & Haefner, J.W. (1998). Parallel implementations of individual-based models in biology. *BioSystems*, 45, 87-97.

Palmer, J., Paton, R.C., & Saunders, J.R. (2004). An extensible individual-based modelling environment using a "space-based" technique on a distributed computing grid. *BioSystems*. Submitted for review on January 2004.

Paton, R.C. (1993). Some computational models at the cellular level. *BioSystems*, 29, 63-75.

Paton, R.C. (1997). The organisations of hereditary information. *BioSystems*, 40, 245-255.

Paton, R.C. & Matsuno, K. (1998). Some common themes for enzymes and verbs. *Acta Biotheoretica*, 46, 131-140.

Paton, R., Staniford, G., & Kendall, G. (1996). Specifying logical agents in cellular hierarchies: Computation. In Cuthbertson, R., Holcombe, M., & Paton, R. (Eds.), *Cellular and molecular biological systems*. Singapore: World Scientific, 105-119.

Rauch, E. (2003). Discrete, amorphous physical models. *International Journal of Theoretical Physics*, 42, 329-348.

RayChaudhuri, D., Gordon, G.S., & Wright, A. (2001). Protein acrobatics and bacterial cell polarity. *Proceedings of the National Academy of Sciences (USA)*, 98, 1322-1334.

Record, Jr., M.T., Reznikoff, W.S., Craig, M.L., McQuade, K.L., & Schlax, P.J. (1996). Escherichia coli RNA polymerase, promoters, and the kinetics of the steps of transcription initiation. In Neidhardt, F.C. (Ed.), *Escherichia coli and Salmonella: Cellular and Molecular Biology*, (2nd ed.), 2. ASM Press: Washington, D.C., 792-821

Richmond, M.H. (1979). "Cells" and "organisms" as habitats for DNA. *Proceedings of the Royal Society of London (Series B)*, 204, 235-250.

Schaff, J. & Loew, L.M. (1999). The virtual cell. *Pacific Symposium on Biocomputing*, 4, 228-239.

Schaff, J., Fink, C.C., Slepchenko, B., Carson, J.H., & Loew, L.M. (1997). A general computational framework for modeling cellular structure and function. *Biophysical Journal*, 73, 1135-1146.

Scheffer, M., Baveco, J.M., DeAngelis, D.L., Rose, K.A., & van Nes, E.H. (1995). Super-Individuals: A simple solution for modelling large populations on an individual basis. *Ecological Modelling*, 80, 161-170.

Shapiro, J.A. (1991). Genomes as smart systems. *Genetica*, 84, 3-4.

Shapiro, J.A. (1992). Natural genetic engineering in evolution. *Genetica*, 86, 99-111.

Thaler, D.S. (1994). The evolution of genetic intelligence. *Science*, 264, 224-225.

Tomita, M., Hashimoto, K., Takahashi, K., Shimizu, T., Matsuzaki, Y., Miyoshi, F., Saito, K., Tanida, S., Yugi, K., Venter, J.C., & Hutchison, C. (1999). E-Cell: Software environment for whole cell simulation. *Bioinformatics*, 15, 72-84.

van Nes, E.H., Scheffer, M., & van den Berg, M.S. (2003). Charisma: A spatial explicit simulation model of submerged macrophytes. *Ecological Modelling*, 159, 103-116.

Way, E.C. (2001). The role of computation in modeling evolution. *BioSystems*, 60, 85-94.

Welch, G.R. (1987). The living cell as an ecosystem: Hierarchical analogy and symmetry. *Trends in Ecology and Evolution*, 2, 305-309.

Zadeh, L.A. (1965). Fuzzy sets. *Information and Control*, 8, 338-353.

Zorman, B., Kapfhammer, G.M., & Roos, R. (2002). Creation and analysis of a JavaSpace-based distributed genetic algorithm. *Proceedings of the International Conference on Parallel and Distributed Processing Techniques and Applications*, 1107-1112.

Chapter VIII

On a Formal Model of the T Cell and Its Biological Feedback

Gabriel Ciobanu, Romanian Academy, Iasi, Romania

ABSTRACT

In this chapter a model of the molecular networks, created by using a network of communicating automata, is described as a dynamic structure, discrete event system, and interesting theoretical results are provided. This formal model provides a detailed approach of the biological system, and its implementation is able to handle large amounts of data. This model is applied to T cell signalling networks. T cell shows a hierarchical organization depending on various factors. Some mechanisms are still unresolved, including contribution of each signalling pathway to each response type. The software tool produced is used to simulate and analyze the T cell behaviour. The simulation reflects, quite faithfully, the process of T cell activation and T cell responses. This increases the confidence in using this model and its implementation both as descriptive and prescriptive tools. The interactions that govern the T cell behaviour are simulated and analyzed, providing statistical correlations according to software experiments, together with new insights on signalling networks that trigger immunological responses. The software tool allows users to systematically perturb and monitor the components of a T cell molecular network, capturing biological information relevant to immunology.

INTRODUCTION

Systems biology is a new holistic view of molecular biology. The challenge for systems biology is to discover answers to molecular biology's questions, and to identify the principles that lead to the actual combination of molecular mechanisms. Today we have long lists of genes and encoded proteins, but simply listing the molecules explains neither how these molecules interact and form molecular networks, nor how the cell processes are triggered and regulated. A useful idea is to add new abstractions, discrete models, and methods able to help our understanding of the biological phenomena, giving us predictive power, useful classifications, possibly new molecular computing paradigms, algorithmic analysis of biomolecular protocols, and a new perspective on the dynamics of the membrane systems.

The understanding of complex molecular networks is difficult due to the great number of molecules, proteins and enzymes, pathways, and feedback loops involved in cellular control. Systems biology aims to have a modular understanding of biological phenomena, focusing on system structure and behaviour analysis. In a systemic way, complex molecular networks can be decomposed into smaller meaningful units; then the process of modelling is understood as a combination of parts along two coordinates: structure and behaviour. This chapter defines a network of communicating automata, including more quantitative details. This approach offers an appropriate framework for modelling and provides considerable theoretical power. It is useful to connect this approach to a general theory of modelling and simulation defined by the discrete event system formalism (Zeigler et al., 2000).

In this chapter, we refer to both the behaviour dynamics and structure dynamics of the T cell signalling network. The system's behaviour is represented as time-varying input/output segments. Given the behaviour of a causal, time-invariant system, we use some abstraction mechanisms to support compositional modelling. Moreover, we consider that our system can change its structure by adapting a rule according to the previous steps. Everything is expressed in a distributed environment making the model compositional, highly adaptable, and realistic. In this way, we can predict unobserved behaviours according to a theoretical framework of discrete-event modelling.

The implementation of the molecular network is provided by a client/server model, well-known in computer networks. This novel software tool is composed of a data server and its clients, together with a graphical tool for a visual representation of network dynamics. It is designed to work over computer networks, using the power of several processors and systems. It is platform-independent and functional over heterogeneous networks.

The results obtained from experiments such as these are processed and interpreted by means of statistics. One of the main goals of statistics is inferring conclusions based solely on a finite number of observations about events possible to happen an indefinite (infinite) number of times. Nonetheless, the strength of conclusions yielded depends on the sample size upon which the analysis is based. In fact, in many cases a contradiction appears between this minimum size required by statistics to make methods applicable and the size that biological wet experiments can provide. This is the reason that correct computer simulators for biological processes are needed; the use of such a tool overcomes the problems of budget, since the cost per experiment (both normal and marginal) is low. Therefore, data sets of desired size can be obtained, thus allowing for correct statistical inferences and hypothesis testing.

The next section of this chapter presents the model and its link to the discrete event system formalism. The following section briefly presents the biology of a T cell network and its behaviour (activation and responses). A section is dedicated to the implementation of our model and experiments for a T cell molecular network. We conclude with the biological relevance of these experiments and the links with molecular computing.

THE MOLECULAR NETWORK AS A NETWORK OF AUTOMATA

The process of modelling and simulation of a system implies three basic objects: the real system, the model, and the simulator. It also implies two main relations: the modelling relation, which ties the real system to the model, and the simulation relation, which connects the model and the abstract simulator (see Example 1).

In this chapter, we refer to the T cell signalling network. The dynamics of molecular networks is provided by a set of communicating automata and appropriate internal transitions for each component of the network or communication steps between components. The approach is systemic; we do not

Example 1.

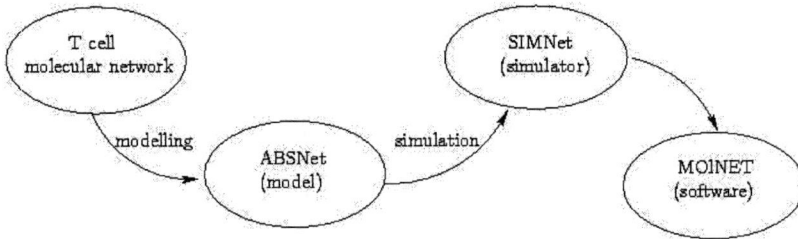

investigate individual automata (molecules), but rather describe their network behaviour both qualitatively and quantitatively.

Generally, for each molecule we define an automaton $M = (X, S, Y, \delta^{int}, \delta^{ext} \lambda, \tau,)$, where

- $X = \{(p, v)| p \in IPorts, v \in Xp\}$, where *IPorts* - input ports, *Xp* - possible input values.
- $Y = \{(p, v)| p \in OPorts, v \in Yp\}$, where *OPorts* - output ports, *Yp* - possible output values.
- S is the *set of states*, with an initial state s_0
- $\delta^{ext} : Q \times X \rightarrow S$ is the *external transition* function, where $Q = \{(s, e)| s \in S, 0 \leq e \leq \tau(s)\}$, and e is the elapsed time since last transition.
- $\delta^{int} : S \rightarrow S$ is the *internal transition* function.
- $\lambda : S \rightarrow Y$ is the *output* function.
- $\tau : S \rightarrow R^{\geq 0}$ is a *time advance* function.

The interpretation for this system is the following: at any time the system is in some state, say s. If no extern event occurs, the system will stay in state s for time $\tau(s)$. Notice that $\tau(s)$ could be a strictly positive real number as one would expect, but it can also take on the values 0 and ∞. In the first case $(\tau(s) = 0)$, the stay in state s is so short that no external event can intervene; we say that s is a *transitory state*. In the second case $(\tau(s) = \infty)$, the system will stay in s forever unless an external event interrupts its slumber; in this case, we say that s is a *passive state*. When the time $\tau(s)$ has elapsed, the system outputs the value $\lambda(s)$ and changes to state $\delta^{int}(s)$. Note that the output is only possible just before internal transitions. If an external event $x \in X$ occurs before this expiration time — that is, when the system is in state (s, e) with $e \leq \tau(s)$ — the system changes to state $\delta^{ext}((s, e), x)$. Thus the internal transition dictates the system's new state when no events occurred since the last transition. The external transition function dictates the system's new state when an external event occurs. In both cases, the system is then in a new state s', with a new elapsed time $\tau(s')$, and the same procedure is repeated.

In order to define a network, we start by considering a subset D of molecules, and for each molecule $i \in D$ we have a set S_i with all possible instances of the i, called states of i. M_i is the automaton for i, I_i is the set of influencers for this molecule i (i.e., possible interaction partners for i), and Z_i is the i-input function $(Z_i: \times_{j \in Ii} Y_j \rightarrow X_i)$.

- The *structure of a network* based on D is defined by $(D, \{M_i\}_{i \in D}, \{I_i\}_{i \in D}, \{Z_i\}_{i \in D})$.
- A change in the network structure $(D, \{M_i\}_{i \in D}, \{I_i\}_{i \in D}, \{Z_i\}_{i \in D})$ is defined as a *change of structure*.

This automata-based approach is similar to the discrete event system specifications for modelling and simulation (Zeigler et al., 2000). The discrete event system specification formalism and its theoretical basis provide a number of important properties such as hierarchy, modular composition, universality, and uniqueness that can support development of simulation models and environments. The support for modelling variable structure systems is provided by the dynamic structure discrete event system specification (DSDEVS) formalism (Barros, 1997). In the context of this theory, the models are of two types: basic and network. A DSDEVS basic model is a discrete event system specification (DEVS) model (Zeigler et al., 2000). The network model is a combination of DSDEVS basic models. *DSDEVS* $N = (\chi, M_\chi)$ defines the DSDEVS dynamic structure network model; χ is the name of the DSDEVS network executive, and $M_\chi = (X_\chi, S_\chi, Y_\chi, \gamma, \Sigma, \delta_\chi, \tau_\chi)$ is the DEVS description for χ, where the network structure Σ, at a state $s_\chi \in S_\chi$, is given by a $\Sigma = \gamma(s_\chi) = (D, \{M_j\}, \{I_j\}, \{Z_j\})$.

A change of structure refers to adding or removing components, or modifying component links. The network structure can be changed when the executive performs an internal or external transition. Since all structural information is kept in the network executive state, any such change will be accompanied by executive transitions. It has been proven that a DSDEVS dynamic structure network model is a DEVS model (Barros, 1997; Zeigler et al., 2000). This condition guarantees that by a hierarchical construction of network models, a basic model is obtained. Any component is a model itself, and because the number of components is variable, it is natural to consider that the model is obtained by coupling the active components. In this way, the model has a variable structure.

The DSDEVS formalism is implemented as a software environment called MOlNET; DSDEVS and MOlNET provide support for building dynamic structure simulation models in an hierarchical and modular manner. This general systems theory-based formalism offers sound semantics for dynamic structure modelling and simulation. The importance of working with a general formal framework for modelling and simulation is given by the fact that, up to a specific modelling relation, we can prove that the simulations reflect faithfully the behaviour of the real system.

We used this approach, and the results were compared with those obtained by experiments. They were quite similar, providing confidence in using the model and its implementation both as descriptive and prescriptive tools.

DSDEVS Basic Model of Molecular Networks

In a basic DSDEVS model for molecular networks, let A be a molecule type. For A we define a *DEVS* that describes the evolution of all type A molecules (complexes), in all the possible states to appear in a molecular network.

Let $M_A = (X_A, S_A, Y_A, , \delta_A^{int}, \delta_A^{ext}, \lambda_A, \tau_A)$ be the DEVS description for A. We consider that a molecule type A can appear in n states in the given molecular network. We assign numbers from 1 to n to these states. Each state can participate in a number of reactions. For each state j of molecule type A, we define a set $R(A,j)$ composed of the rules that contain the molecule A in state j. Let $r = aA_j + bB_k \rightarrow cC + dD$ be such a reaction. We denote by A_j the fact that A is in state j. In Table 1 we give the full description of the states and transition functions for A_j with respect to a reaction r, and in Figure 1 we describe the transition graph for a reaction.

For each state j of the molecule type A, we define a set $X_{A,j} = \cup_{r \in R(A,j)} \{(X_{A,j}, (cod, p_r))\}$ where $cod \in \{ok, no, 1\}$, and p_r is the interaction partner of A_j regarding the reaction r. We have $X_A = X_{A,1} \times \ldots \times X_{A,n}$. In a similar way $Y_{A,j} = \cup_{r \in R(A,j)} \{(Y_{A,j}, (cod, A_j, p_r)), (Y_{A,0}, (cod, r, A_j))\}$, where $cod \in \{ok, no, 1\}$ and p_r is the partner of A_j regarding the reaction r. We consider $Y_A = Y_{A,1} \times \ldots \times Y_{A,n}$.

Messages of type $(Y_{A,0}, (cod, r, A_j))$ are sent to the executive. We denote by $S_{A,j,r} = \{s_i | 1 \leq i \leq 8\}$ the set of states specific to the molecule A in state j when it participates to a reaction r. $S_{A,j} = \cup_{r \in R(A,j)} S_{A,j,r}$ is the set of states for molecule A when it is in state j. Using these sets we define $S_A = S_{A,1} \times \ldots S_{A,n}$, the set

Table 1. The states and the transition functions for a molecule A *appearing in state* j *in a reaction* r *(the states* $s_1, s_2, s_3,$ *and* s_4 *are active states, whereas* $s_5, s_6, s_7,$ *and* s_8 *are passive states;* tr *is the reaction time for* r, *and* x *is a molecule of* B_k*)*

states ($S_{A,j,r}$)	$\delta_{A,j}^{int}$	$\delta_{A,j}^{ext}$	$\tau_{A,j}$	$\lambda_{A,j}$
$S_1 = (c, 0, B_k)$	s_2		0	$(Y_{A,j}, (1, A_j, B_k))$
$S_2 = (c, 1, B_k)$		$(s_2, (X_{A,j}, (ok, B_k))) \rightarrow s_3$ $(s_2, (X_{A,j}, (n_o, B_k))) \rightarrow s_4$	∞	
$S_3 = (c, 2, B_k)$	s_0		tr	$(Y_{A,0}, (ok, r, A_j))$
$S_4 = (c, 3, B_k)$	s_0		0	$(Y_{A,0}, (no, r, A_j))$
$S_5 = (c, 4, B_k)$		$(s_5, (X_{A,j}, (1, B_k))) \rightarrow s_6$ $(s_5, (X_{A,j}, (1, x))) \rightarrow s_7$	∞	
$S_6 = (c, 5, B_k)$	s_8		0	$(Y_{A,j}, (ok, A_j, B_k))$
$S_7 = (c, 6, B_k)$	s_5		0	$(Y_{A,j}, (no, A_j, B_k))$
$S_8 = (c, 7, B_k)$	s_0		tr	$(Y_{A,0}, (ok, r, A_j))$

of states for the DEVS description M_A corresponding to the molecule type A. Considering the defined functions $\delta_{A,j}^{int} : S_{A,j} \rightarrow S_A$, $1 \leq j \leq n$, we define $\delta_A^{int} : S_A \rightarrow S_A$ as follows: $\forall s \in S_A$, $s = (s_1, \ldots, s_n)$, $s_j \in S_{A,j}$, $1 \leq j \leq n$, $\delta_A^{int}(s) = (\delta_{A,1}^{int}(s_1), \ldots, \delta_{A,n}^{int}(s_n))$. The functions δ_A^{ext} and λ_A are defined in a similar manner. Let s be in S_A, with $s = (s_1, \ldots, s_n)$. For each $s \in S_A$ with $s = (s_1, \ldots, s_n)$, we define $T_s = (t_{1,s}, \ldots, t_{n,s})$. If $s' = (s_1, \ldots, s_j', \ldots s_n)$ with $\delta_{A,j}^{int}(s_j) = s_j'$ or $\delta_{A,j}^{ext}(X_{A,j}, s_j) = s_j'$, then $T_s'' = (t_{1,s} - \tau_A(s), \ldots, \tau_{A,j}(s_j), \ldots, t_{n,s} - \tau_A(s))$ where $\tau_A(s) = min(t_{1,s}, \ldots, t_{n,s})$. For the initial state $T_{s0} = (\tau_{A,1}(s_{0,1}), \ldots, \tau_{A,n}(s_{0,n}))$.

DSDEVS Network Model of Molecular Networks

In order to define a network model, we start by considering a subset D of molecules. For all $A \in D$, M_A the system described earlier applies for the molecule A, with $I(A)$ as the set of possible partners for A, and Z_A the A-input function ($Z_A : \times_{B \in I(A)} Y_B \rightarrow X_A$). Let $Z_{A,B}$ be the projection of Z_A onto component Y_B, with $Z_{A,B}(y) = (X_{A,h}, (cod, exp))$ if k_h of y is A, and $Z_{A,j}(y) = \emptyset$ if k_h of y is not A, where $y = (cod, exp, k_h) \in Y_j$.

The clients send information to the executive about the reactions that take place, or about unsuccessful attempts to react. When quantitative changes appear in the network, the executive recomputes the putative times for each reaction (according to Gibson's algorithm), and modifies the clients' structures such that for the next reaction the selected client to participate is the one with

Figure 1. The transition graph for a reaction; if the system passes from s_0 to s_1, then A is the active reactant, but if the system passes from s_0 to s_5, A is the passive reactant (this is decided by the executive)

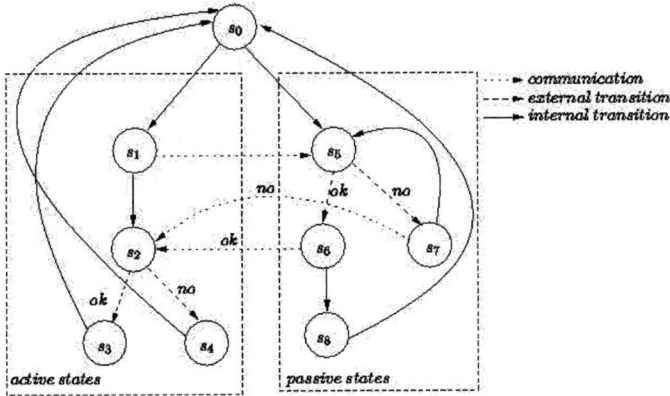

the least putative time. This process of the executive can be described as follows.

When receiving a message $(code, r, A_j)$,

if *code* = ok
> change the system structure to reflect the quantitative aspects of the reaction r;
> recalculate the putative time;
> r_1 = reaction with the least putative time and A in some state k appears in r_1;

if *code* = no
> r_1 = reaction with the least putative time, with A in some state k appearing in r_1, and $r_1 \neq r$.

the executive modifies the structure of A:

- $d_{A,k}(s_0) = s_1, s_1 \in S_{A,j,k}$ if A is the active reactant.
- $d_{A,k}(s_0) = s_5, s_5 \in S_{A,j,k}$ if A is the passive reactant.

Considering a simplified version of the dynamic structure network model, we prove that it has the same computational power as Turing machines

(Ciobanu et al., 2003). We have improved the previous results presented by Ciobanu and Huzum (2003), and provide the following new result: a dynamic structure network model with only two clients is Turing complete; that is, it is able to compute all of the computable functions. The proofs use some results and techniques of the formal language theory (Rozenberg & Salomaa, 1997). We use matrix grammar with appearance checking in a binary normal form, and some technical results from Dassow and Păun (1989). The results are interesting and inspiring, and the process recommends itself for breadth and depth of the philosophical questions regarding the computational power of the biochemical networks and the possibility of designing molecular computers.

SIGNALLING PATHWAYS AND T CELL ACTIVATION

Many biological activities depend on the ability of proteins to communicate specifically with each other and with other molecules. This is particularly obvious for the signalling pathways that operate in living cells. These signalling pathways allow the cell to receive, process, and respond to various *signals* from its environment. The signals are molecules (hormones, neurotransmitters, cytokines) that indicate the beginning and the termination of one or more intracellular processes. Signal *transduction* refers to the process by which the signals are transmitted via receptors to the interior of the cell through the signalling pathways. Classically, a linear signal transduction pathway can be described as follows. A first messenger molecule (the signal) outside the cell is sensed at the cell membrane by a receptor. Receptor binding is transduced into an intracellular event by activation of enzymes that synthesize second messenger molecules. These molecules promote covalent modifications (by protein kinase or protein phosphatase activities) and allosteric regulation of other intracellular proteins that ultimately cause specific changes in gene expression. In living cells, these linear signalling pathways cross each other and form a network of interactions. This signalling network has several emergent properties that the individual pathways do not have. We consider the signalling network that grounds T cell activation as a case study of our model.

T cells play an important role in orchestrating the immune responses to foreign aggressors. When a T cell recognizes a foreign antigen, it initiates several signalling pathways and the cell activates. The recognition of foreign antigens is an extremely *sensitive, specific,* and *reliable* process, and models for T cell signalling are necessary to understand how these properties arise. So

Figure 2. TCR signalling pathways

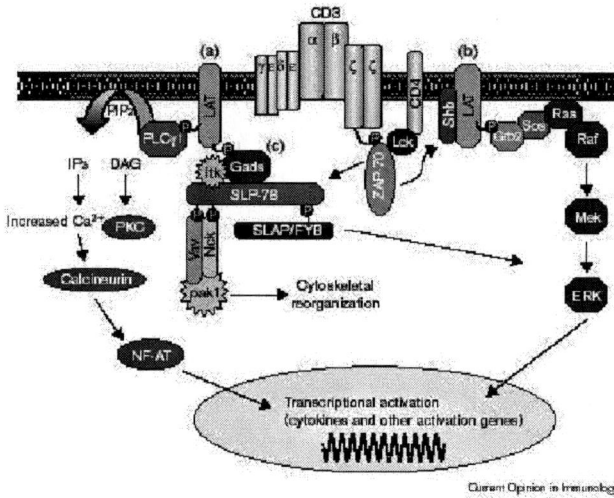

Current Opinion in Immunology

far, the study of T cell activation has benefited from the use of mathematical models (Chan, 2002). The key event for T cell activation is an appropriate interaction between the T cells armed with T cell antigen receptors (TCR) and the professional antigen presenting cells (APC).

TCR only recognize foreign antigens in the form of short peptides presented in the groove of a molecule on the surface of the APC, known as the major histocompatibility complex (MHC). The physical interaction of TCR with MHC-peptide complexes is unique among signalling systems in that it takes place over a continuum of binding values. The recognition of antigens initiates signal transduction. This process can be broken down into a series of discrete steps that are related to molecular events (interactions, state transitions) within the signalling pathways. These steps are shown in Figure 2, reprinted from Myung et al. (2000). T cell responses show a hierarchical organization depending on the extent of TCR occupancy, the duration of antigen binding, the timing of encounters, and the engagement of costimulatory receptors. TCR is a very complex structure composed of a minimum of eight strongly associated chains. The actual arrangement and stoichiometry of CD3 and TCR chains within TCR complex is still unknown. We refer here only to a part of the signalling network, namely the activation ZAP-70 and phosphorylation of the adapters LAT, which are essential for connecting to the major intracellular signalling pathways $Ca+2$/calcineurin and Ras/Rac/MAPK kinases. In this sense, although many other receptor interactions may contribute positively or negatively to the quality and the quantity of the T cell immune responses, TCR signalling upon antigen recognition determines whether or not a response of any type will result (Ciobanu et al., 2002).

Model Evolution and Behaviour

In order to illustrate the evolution of the DSDEVS, we consider the reaction between ZAP-70 and LAT followed by the binding of $GADS \oplus SLP$-76 to phosphorylated LAT.

As we can see in Figure 3, first there is some communication between ZAP-70 and LAT followed by a transition of these two systems in corresponding states s_3 and (s_8, s_0). The systems wait in these states for the time indicated by their reaction, and then both send a message to the executive. The executive modifies the structure of the systems to reflect the quantitative aspect of the reaction. The executive also modifies the transition function from the initial state for these two systems, allowing the binding of the complex $GADS \oplus SLP$-76 to phosphorylated LAT.

We give a more detailed description of the state transition and communication between systems for the reaction between ZAP-70 and LAT. The network connection between the system M_{ZAP-70} and M_{LAT} is presented in Figure 4. Each system is represented by a rectangle (e.g., LAT), and the communication channels are represented by arrows. For each state of a molecule (e.g., $LAT_{0,0,0,0}$), there is a pair of communication ports: one for input and another for output (e.g., $X_{LAT,1}$ and $Y_{LAT,1}$). In order to avoid message collision, a function for each system is defined (e.g., Z_{LAT}); these functions are represented by circles.

We consider the following reaction formula:

$$\text{a} ZAP\text{-}70_{?,?,1} + \text{b} LAT_{0,0,0,0} \rightarrow \text{a}_1\, ZAP\text{-}70_{?,?,1} + \text{b}_1\, LAT_{1,1,1,1}$$

where indices of the above elements represent domain configuration.

The evolution of the system M_{ZAP-70} can be described as follows: the initial state for M_{ZAP-70} is $s_0 = (c, 0, none)$. This is a transitory state, and from this state the system goes to state $s_1 = (c, 0, LAT_{0,0,0,0})$, also a transitory state. We use this state to send a message to $LAT_{0,0,0,0}$. The next state is $s_2 = (c, 1, LAT_{0,0,0,0})$; the system remains in this state until an external event interrupts its slumber. The external event consists of an answer from $LAT_{0,0,0,0}$; if the answer is $(ok, LAT_{0,0,0,0})$, then the system goes to state $s_3 = (c, 2, LAT_{0,0,0,0})$. The system will stay in this state for a time equal to the reaction time tr. When the time has

Figure 3. State transition graph

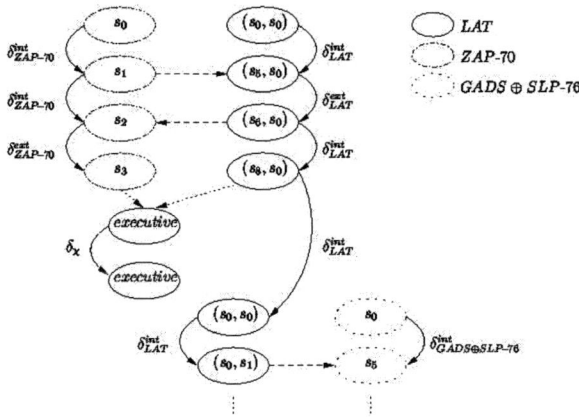

Figure 4. The network connection between the system M_{ZAP-70} *and* M_{LAT}

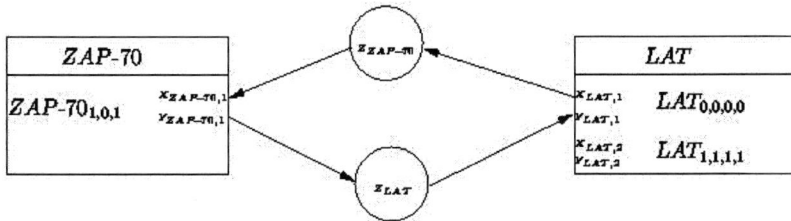

elapsed, a message is sent to the executive and M_{ZAP-70} goes to its initial state s_0.

IMPLEMENTATION

Molecular networks and computer networks look and behave similarly. In this context, a valuable approach to model and simulate the molecular networks is to develop a software system that runs over computer networks. We developed a novel software system called MOlNET as an implementation of our DSDEVS model for molecular networks. Like the DSDEVS model, MOlNET has two main entities: the data server and the complex clients. The data server is the implementation of the executive, and the complex clients implement the basic DSDEVS model, simulating molecule complexes. MOlNET implements also the communication between the clients that describe molecule complexes. The DSDEVS theoretical framework ensures model validity and simulator correctness. A simulator performs the implicit operations of the DSDEVS models. It is able to perform simulations contained in both basic models and dynamic structure network models. Considering the simulation relations, we can prove that every basic DSDEVS model for molecular networks is simulated by MOlNET *clients*, the executive of the DSDEVS model for molecular networks is simulated by our *data server*, and the communication between entities of the DSDEVS model for molecular networks is simulated by MOlNET communication protocols.

For implementation, we used C programming language, BSD-socket interface for network communication, and GTK 2.0 for the graphical interface. The entire architecture is shown in Figure 5. The software has a modular architecture, which allows us to easily integrate other tools, or even to use various interaction algorithms:

- *The graphic server:* It ensures a friendly graphical interface to the user. The user is provided with multiple facilities such as saving and loading simulations and a graphical interface for both input and output data, modifying the system data while simulations run or defining tracers to provide charts for the quantitative evolution of different complex clients of the system. The graphic server is also responsible for distributing the processes over the available hosts by communicating with the generator client through a specific protocol.
- *The data server:* This server is the correspondent of the executive in DEVS. Its main role is to provide the clients with data regarding the interaction partners.
- *The generator client:* This is a process responsible only for starting other processes (i.e., the data server and client processes).
- *The clients:* We have two types of clients — complex clients and membrane clients. Each complex client represents a type of molecule complex. Since the behaviour of a molecule complex is given by the domain configuration of its proteins, each configuration can be viewed as a different complex. According to the interaction algorithm, each client initiates reactions with its interaction partners and the reaction results. In this way, we avoid involving any other process in the interactions. The membrane clients simulate membranes, keeping track of the amounts of all

Figure 5. MOlNET architecture

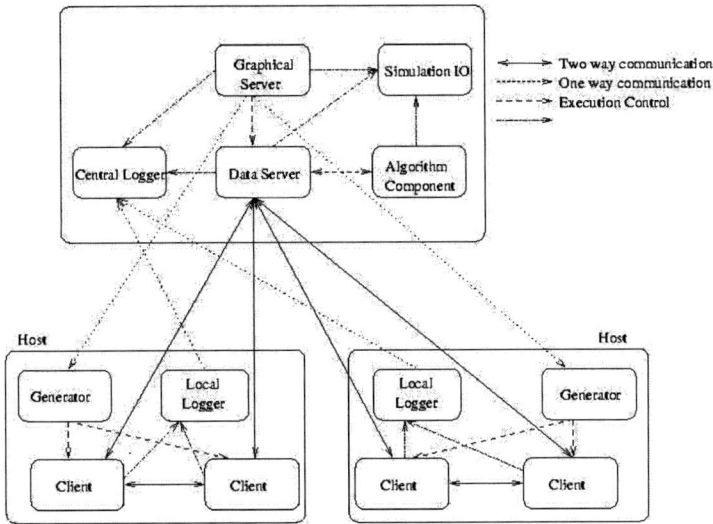

the molecule complexes that lie within. For each particular complex that appears in a tracer, its parent membrane informs the graphic server about the amount of modifications of that particular complex.

The output data consist of tracers. Each tracer is given a group of molecule complexes, and it provides both an output file and a visual representation (i.e., a chart) containing the evolution in time of the given complexes.

The user is provided with an intuitive graphical interface for introducing the input data for a simulation. The input data window includes seven sections, corresponding to molecules, molecule complexes, membranes, system initializing, rules, membrane rules, and hosts. A snapshot is given in Figure 6.

This implementation uses computer networks for distributing the computation, providing the advantage of executing simulations over large number of molecules, and so providing valuable and relevant data about the behaviour of particular molecular networks. Generally speaking, a good simulation of T cells activity opens new perspectives in cancer immunology, treatment of autoimmune disorders, and transplantation management. We simulate and analyze the interactions that drive T cell behaviour by using computer networks and discrete events system specifications. The MOlNET experiments provide data that could be interpreted with statistical methods. These experiments could

Figure 6. A screenshot of MOlNET software

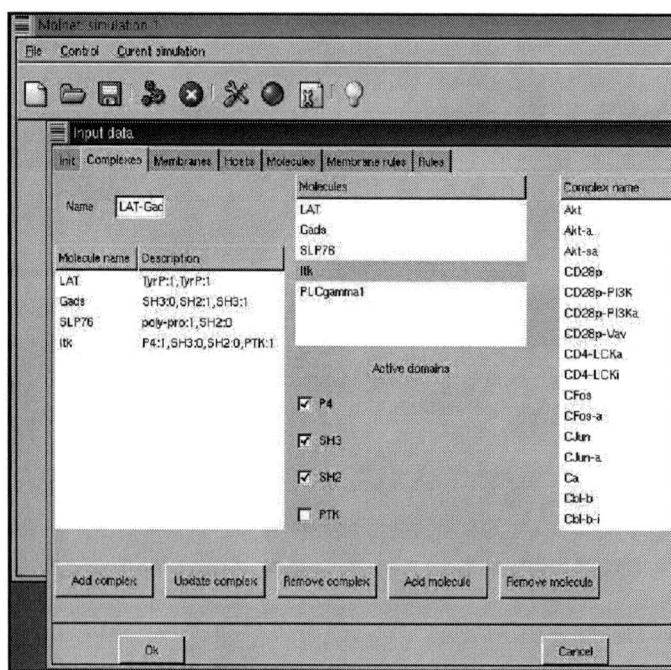

provide a way to systematically perturb and monitor the signalling network outcomes by adding or deleting new molecules, modifying the corresponding amounts and rates, establishing new interactions between components, or wiping out the existing interactions due to mutations. In this way, we can represent various factors (T cell type, the number of triggered TCR, the presence or absence of costimulation) that play a role in determining the outcome of the T cell. The simulation can contribute to explaining how these factors determine differences in the formation and composition of the TCR signalling complexes, and how they drive various biological consequences of T cell activation.

MO1NET EXPERIMENTS AND THEIR BIOLOGICAL RELEVANCE

Biological behaviour is strongly influenced by the ability of molecules to communicate specifically with each other within large molecular networks.

Crucial for T cell behaviour is the signalling network that could engage various cell responses due to potentially different signal types, quantities, and durations. Here we describe the network behaviour both qualitatively and quantitatively.

Many studies on T cell biology have revealed different types of functional responses (activation, proliferation, anergy, cell death) to differences in TCR stimulation. It is known that TCR engagement under some circumstances leads to proliferation and effector function, whereas under other conditions TCR stimulation leads to unresponsiveness (anergy). The variables that shape the response to antigens are the concentration of antigens, the duration of antigen binding, the timing of encounters, and the engagement of other receptors (such as CD28 or CTLA4). These T cell responses can be broken down into a series of discrete steps that are related to molecular events within a larger molecular network. Recent data suggest that unbalanced activation of NFAT relative to its cooperating AP-1 imposes the genetic program of T cell anergy that opposes the program of activation-proliferation mediated by NFAT-AP1 complex. Based on these results, we simulated the molecular network that could drive full T cell activation or T cell anergy. We ran some experiments with the MOlNET simulator, using different input amounts for TCR and CD28 such as their ratio to vary across an interval. The output data of interest were the ratio between NFAT and AP1, the proteins with a main role in deciding the T cell behaviour. In Figure 7 represents data yielded by experiments.

We varied inputs over the interval $10^3 \ldots 10^5$ for the number of molecules per each molecular species, and $10^{-3} \ldots 10^{-5}$ for the Michaelis-Menten constant of enzymatic reactions. However, these conditions may not always hold true and further restrictions may be imposed. On the other hand, they could reflect the diversity of molecular environments throughout T cell population. To be very rigorous, each reaction may be varied in terms of the molecular number of molecules that participate and of kinetic constants. Varying these parameters for all reactions in the network produces an enormous number of MOlNET experiments.

We applied the bootstrap method in the case of ratio between amounts of NFAT and AP1 at the end of experiments made with MOlNET. After generating 500,000 bootstrap samples from the original ratio sample, we obtained an estimate of the mean equal to 0.8023, as compared to the sample average of 0.8071. The full range of the bootstrap estimates was (0.2140, 2.4179) and from it we were able to indicate the 95 percent confidence interval for the mean to be (0.7071, 1.9003). For comparison, the 95 percent confidence interval obtained by using the classical t-test was (0.7220, 0.8921),

Figure 7. Data yielded by the MOlNET simulator from experiments regarding T cell activation; on the X axis experiments are represented in the order they were conducted (it can be seen that ratios TCR/CD28 and NFAT/AP1 have similar trends)

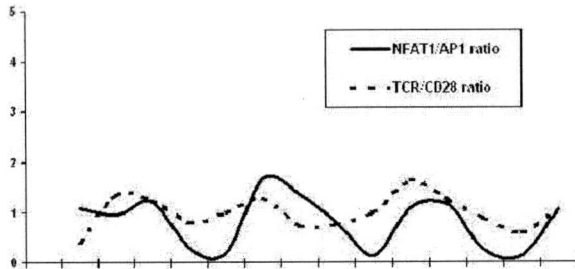

which shows that the population of ratios has a distribution very close to normal. Also, by applying the bootstrap method, we obtained a similar conclusion with regard to the ratio TCR/CD28.

Other experiments we made with MOlNET are related to the Lck recruitment model, which involves successive activations and inactivation of ZAP-70, Lck, and phosphatase. We ran several tests regarding the quantitative aspects of this particular interaction network. A chart representing the distribution of quantitative modification corresponding to each of the substances involved is presented below (we used 1.2 for a 20 percent increase). The samples follow a normal distribution, with the mean falling on either side of the 1.0 mark, depending on the evolution of the system. For example, the mean for inactivated Lck is clearly less than 1, while the mean for its counterpart activated Lck is greater than 1.

Furthermore, based on the obtained data, we could predict the 95 percent confidence interval for the mean of activated Lck modification as being the interval (0.91; 1.14). In this way, the data yielded by using MOlNET can be analyzed in order to find statistical correlations of the mechanisms inside the cell; it is also possible to make predictions of the network dynamics.

To summarize, in this chapter we emphasized the importance of simulations based on some mathematical theory in order to understand the complex molecular networks systematically. The chapter shows that starting from a network of automata and taking useful details into consideration, we get a

Figure 8. Distribution of quantitative modifications in the Lck recruitment

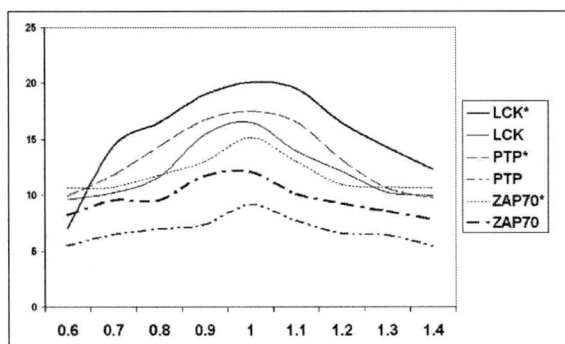

formal approach similar to an existing formal framework for modelling and simulation. Together with the network client/server model, our approach leads to a novel software and provides new insights in molecular networks. We present an implementation for this model and analyze the interactions that influence the T cell behaviour by using statistics correlations of the software experiment results. As a result, we get relevant biological information on T cell behaviour, particularly on T cell responses. The simulations explain how various factors play a role in determining the T cell response, providing some corresponding thresholds connecting input and output values of T cell mechanism. As far as we know, the existence of a coordinator (executive) for the molecular interaction is a new feature; such a coordinator was recently advanced by some biologists. Therefore, it is difficult to compare the architecture of our system with other systems. Regarding its functionality, we can mention its flexibility, modularity, and expressive power. The current approach may serve as a platform for further experimental and theoretical investigations.

ACKNOWLEDGMENTS

Many thanks to the anonymous referees for their useful remarks and comments. This work was partially supported by an academic research grant of the National University of Singapore (NUS). A number of students have contributed during their internship at NUS, including Bogdan Tanasă, who has contributed to the description of T cell signalling pathways, Dorin Huzum and Gabriel Moruz, who have contributed to the implementation, and Daniel Dumitriu in statistical interpretation.

REFERENCES

Barros, F. J. (1997). Modeling formalisms for dynamic structure systems. *ACM Transactions on Modeling and Computer Simulation*, 7(4), 501-515.

Chan, C. (2002). *Modeling T cell activation*. PhD thesis, Center for Nonlinear Dynamics and its Applications, University College, London.

Ciobanu, G. & Huzum, D. (2003). Discrete event systems and client-server model for signaling mechanisms. In Priami, C. (Ed.), *Computational methods in systems biology. Lecture Notes in Computer Science*, 2602, 175-177.

Ciobanu, G., Dumitriu, D., Huzum, D., Moruz, G., & Tanasă, B. (2003). Client-server P systems in modeling molecular interaction. In Păun, Gh., Rozenberg, G., Salomaa, A., & Zandron, C. (Eds.), *Membrane computing. Lecture Notes in Computer Science*, 2597, 203-218.

Ciobanu, G., Tanasă, B., Dumitriu, D., Huzum, D., & Moruz, G. (2002). Simulation and prediction of T cell responses. *Proceedings of the 3rd Conference on Systems Biology ICSB'02*, Stockholm, 88-89.

Dassow, J. & Păun, Gh. (1989). *Regulated rewriting in formal language theory*. Berlin: Springer.

Myung, P., Boerthe, N., & Koretzky, G. (2000). Adapter proteins in lymphocyte antigen-receptor signalling. *Current Opinion in Immunology*, 12, 256-266.

Rozenberg, G. & Salomaa, A. (1997). *Handbook of formal languages*. Berlin: Springer.

Zeigler, B. P., Praehofer, H., & Kim, T. G. (2000). *Theory of modeling and simulation*. Orlando, FL: Academic Press.

Chapter IX

Formal Modelling of the Dynamic Behaviour of Biology-Inspired, Agent-Based Systems

Petros Kefalas, CITY College, Thessaloniki, Greece

George Eleftherakis, CITY College, Thessaloniki, Greece

Mike Holcombe, University of Sheffield, United Kingdom

Ioanna Stamatopoulou, South-East European Research Centre, Thessaloniki, Greece

ABSTRACT

Multi-agent systems are highly dynamic since the agents' abilities and the system configuration often changes over time. In some ways, such multi-agent systems seem to behave like biological processes; new agents appear in the system, some others cease to exist, and communication between agents changes. One of the challenges is to attempt to formally model the dynamic configuration of multi-agent systems. Towards this aim, we present a formal method, namely X-machines, that can be used to formally specify, verify, and test individual agents. In addition, communicating X-machines provide a mechanism for allowing agents to communicate messages to each other. We utilize concepts from biological processes in order to identify and define a set of operations that are able to reconfigure a multi-agent system. In this chapter we present an example in which a biology-inspired system is incrementally built in order to meet our objective.

INTRODUCTION

An *agent* is an encapsulated computer system that is situated in some environment and is capable of flexible, autonomous action in that environment in order to meet its design objectives (Jennings, 2000). The extreme complexity of agent systems is due to substantial differences in attributes between their components, high computational power required for the processes within these components, huge volume of data manipulated by these processes, and the possibly excessive amount of communication needed to achieve coordination and collaboration. The use of a computational framework that is capable of modelling both the dynamic aspect (i.e., the continuous change of agents' states together with their communication) and the static aspect (i.e., the amount of knowledge and information available) will facilitate modelling and simulation of such complex systems.

The multi-agent paradigm can be further extended to include biology-inspired systems. Many biological processes seem to behave like multi-agent systems, as, for example, a colony of ants or bees, a flock of birds, or tissue cells (Dorigo et al., 1996). The vast majority of computational biological models are based on an assumed, fixed system structure that is not realistic. The concept of growth, division, and differentiation of individual components (agents) and the communication between them should be addressed to create a complete biological system that is based on rules that are linked to the underlying biological mechanisms allowing the dynamic evolution.

For example, consider the case of a tissue, which consists of a number of cells. Each cell has its own evolution rules that allow it to grow, reproduce, and die over time or under other specific circumstances. The cells are arranged in a two- or three-dimensional space, and this layout dictates the way cells interact with others in the local neighbourhood. The rules of communication and exchange of messages are dependent on the particular system. Being a dynamic system, the structure of the tissue — that is, the configuration of cells and therefore their interaction — may change over time, thus imposing a change in the global state of the system. New cells are born, others die, and some are attached to or detached from existing ones (Figure 1).

In the last years attempts have been made to devise computational models in the form of generative devices (Păun, 2000, 2002; Banatre & LeMetayer, 1990; Adleman, 1994; Holcombe, 2001). In this chapter we focus on how we can model such dynamic multi-agent systems, taking as a principle example a biology-inspired system like the one we have discussed, by using a formal method called X-machines. A particular type of X-machines, called communicating X-machines (Kefalas et al., 2003b), will be used. X-machines and communicating X-machines have been successfully used in the past for the modelling of stand-alone as well as complex communicating systems (Kefalas et al., 2003c). We consider this formal method, not only as a theoretical tool for specification, but also as a practical tool for modelling that eventually leads to the implementation of correct systems. The important aspect of using X-machines as a modelling method is that they support formal verification (Eleftherakis & Kefalas, 2001) and complete testing (Holcombe & Ipate, 1998). Thus, formal models expressed as X-machines can be checked to see if they satisfy given properties through model checking, and the potential implementation can be tested against the model to identify errors.

At least three benefits of using a communicating X-machine for modelling multi-agent systems are emphasized in this chapter:

- It is a natural environment for simulating dynamic systems.
- It is flexible enough so that when a new component is introduced into the play, the others need not be restructured (this feature will become useful when simulating cell division, death, etc.).
- It is much more efficient than single X-machines initially used to simulate such systems.

Figure 1. The dynamic evolution of a tissue consisting of individual cells

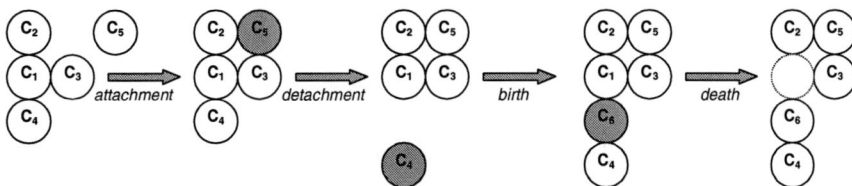

This chapter will mainly concentrate on X-machines since it has been demonstrated that some variants of X-machines may become suitable for molecular computing models due to their completeness property, their capability of naturally simulating different models, their suitability to define dynamic systems reacting to input stimuli, their flexibility in expressing hierarchical or distributed systems, and their ability to capture hybrid specifications.

Our objectives are:

- To demonstrate that X-machines are suitable for modelling individual agents by representing their internal data and knowledge and how the stimuli received from the environment could change their internal state.
- To present the way in which communicating X-machines can facilitate modelling of multi-agent systems, but also their inadequacy to deal with the evolution of such systems.
- To identify and define a number of meta-operations that will enable the dynamic change of the configuration of multi-agent systems.
- To suggest how dynamic multi-agent systems can be modeled as a whole.

One may argue that modelling multi-agent systems may require a new variant of X-machines that maps the particular requirements for such systems. However, the rationale behind using X-machines with their original definition is to be able to exploit, to the maximum possible extent, the legacy of X-machine theory concerning complete testing and formal verification. The particular requirements of multi-agent systems are dealt at a meta-level, thus not affecting modelling, testing, and verification of the individual agent models.

The first section of this chapter presents a background on formal methods and their use in agent-based as well as biology-inspired systems. The chapter goes on to provide an analytical description of X-machines, accompanied by the appropriate definitions and examples. Next, a number of definitions for operators, namely *COM, GEN, ATT, DET* and *DES*, that could handle the dynamic structure of a multi-gent system are given. Also, two approaches for controlling the dynamic behaviour are discussed. Finally, the chapter concludes with a discussion on the benefits of using X-machines as a modelling tool.

BACKGROUND

A variety of formal methods have been used for software system specification. Specification has centered on the use of models of data types, either functional or relational models such as *Z* (Spivey, 1989) or *VDM* (Jones,

1990), or axiomatic ones such as *OBJ* (Futatsugi et al., 1985). Although these models have led to some considerable advances in software design, they lack the ability to express the dynamics of the system. Also, transforming an implicit formal description into an effective working system is not straightforward. Other formal methods, such as finite state machines (Wulf et al., 1981) or Petri nets (Reisig, 1985), capture the essential feature, which is "change", but fail to describe the system completely, since there is little or no reference at all to the internal data and how these data are affected by each operation in the state transition diagram. Other methods, such as statecharts (Harel, 1987), capture the requirements of dynamic behaviour and modelling of data but are rather informal with respect to clarity and semantics.

In agent-oriented engineering, there have been several attempts to use formal methods, each one focusing on different aspects of agent systems development. One of them was to formalise PRS (Procedural Reasoning System), a variant of the BDI architecture (Rao & Georgeff, 1995) with the use of Z, in order to understand an agent's architecture in a better way, to be able to move to the implementation through refinement of the specification and to develop proof theories for the architecture (d'Iverno et al. 1998). Trying to capture the dynamics of an agent system, Rosenschein and Kaebling (1995) viewed an agent as a situated automaton that generates a mapping from inputs to outputs, mediated by its internal state. Brazier et al. (1995) developed the DESIRE framework, which focuses on the specification of the dynamics of the reasoning and acting behaviour of multi-agent systems. In an attempt to verify whether properties of agent models are true, work has been done on model checking of multi-agent systems with re-use of existing technology and tools (Benerecetti et al., 1999; Rao & Georgeff, 1993). Towards implementation of agent systems, Attoui and Hasbani (1997) focused on program generation of reactive systems through a formal transformation process. A wider approach is taken by Fisher and Wooldridge (1997) who utilise Concurrent METATEM to formally specify multi-agent systems and directly execute the specification, while verifying important temporal properties of the system. Finally, in a less formal approach, extensions to UML to accommodate the distinctive requirements of agents (AUML) were proposed by Odell et al (2000).

In recent years, some of the above formal approaches have been considered for modelling different biological phenomena and aspects of living organisms. Others have attempted to devise new computational models, such as P Systems (Păun, 2000), originating their inspiration from biochemical processes that occur in living cells (Università, 2003; Leiden University, 2003) or Gamma (Banatre & Le Metayer, 1990) and Cham (Berry & Boudol, 1992) by modelling the concurrent behaviour of systems. Computational models such as Boolean networks (Kauffman, 1993) have also been developed to express biochemical reactions occurring at the cell level, or X-machines, utilized to model metabolic pathways (Holcombe, 1988) and the behaviour of bee and ant colonies (Gheorghe et al., 2001; Kefalas et al., 2003c). Hybrid approaches have been developed to facilitate the specification of complex aspects related to micro or macro biostructures (Duan et al., 1995; Balanescu et al., 2002; Gheorghe et al., 2003). And finally, attempts have been made to show the equivalence of new computational models (Kefalas et al., 2003a).

MODELLING WITH COMMUNICATING X-MACHINES

X-Machines

An *X-machine* is a general computational machine introduced by Eilenberg (1974) and extended by Holcombe (1988) that resembles a finite state machine (FSM) but with two significant differences:

- A memory is attached to the machine.
- The transitions are not labeled with simple inputs but with functions that operate on inputs and memory values.

These differences allow X-machines to be more expressive and flexible than the FSM. Other machine models such as pushdown automata or *Turing machines* are too low-level and are therefore of little use for specification of real systems. X-machines employ a diagrammatic approach for modelling the control by extending the expressive power of the FSM. They are capable of modelling both the data and the control of a system. Data are held in the memory structure. Transitions between states are performed through the application of functions, which are written in a formal notation and model the processing of the data. Functions receive input symbols and memory values and produce output while modifying the memory values (Figure 2). The machine, depending on the current state of control and the current values of the memory, consumes an input symbol from the input stream and determines the next state, the new memory state, and the output symbol, which will be part of the output stream.

Figure 2. An abstract example of an X-machine; φ_i: functions operating on inputs and memory, S_i: states — The general format of functions is φ $(\sigma, m) = (\gamma, m')$

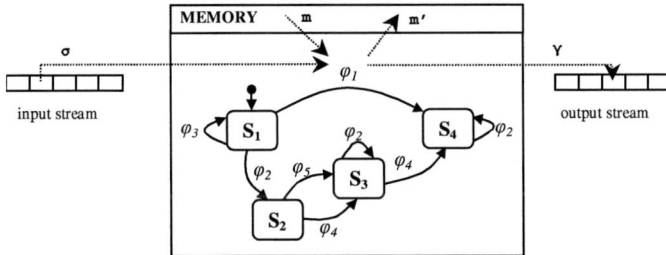

Definition. A *deterministic stream X-machine* (Holcombe & Ipate, 1998) is an 8-tuple $X = (\Sigma, \Gamma, Q, M, \Phi, F, q_0, m_0)$, where:

- Σ, Γ are the input and output finite alphabets, respectively.
- Q is the finite set of states.
- M is the (possibly) infinite set called memory.
- Φ represents the type of the machine X, a finite set of partial functions φ that map an input and a memory state to an output and a new memory state, $\varphi: \Sigma \times M \to \Gamma \times M$.
- F is the next state partial function, $F: Q \times \Phi \to Q$, that given a state and a function from the type Φ, denotes the next state. F is often described as a state transition diagram.
- q_0 and m_0 are the initial state and initial memory, respectively.

The X-machine integrates both the control and data processing while allowing them to be described separately. The X-machine formal method forms the basis for a specification or modelling language with a great potential value to software engineers. It is rather intuitive, and formal descriptions of data types and functions can be written in any known mathematical notation. Finally, X-machines can be extended by adding new features to the original model, such as hierarchical decomposition and communication, which will be described later. Such features are particularly interesting in agent-based systems.

Modelling a Cell Agent as an X-Machine

In this section, we discuss the modelling of an agent as a stream X-machine. The model we present here is simple for reasons of exposition and it follows various underlying assumptions that entail an overall simplicity (e.g., discretisation of possible movements and directions, identical cells, two-dimensional plane, etc). Consider a simple agent, such as a cell that is born, matures, and eventually dies. Each of these life states can be described in terms of a state of an X-machine. While the agent is mature it is able to reproduce. The agent can move in four possible directions if it is forced to do so (i.e., by receiving an appropriate input). Naturally, the agent may exhibit other behaviours as well, but for the sake of simplicity, we will omit those. The interested reader may refer to Kefalas (2002) for further details. A state transition occurs when the appropriate input is received. The next state transition diagram, represented by F, is depicted in Figure 3. The definition of the cell formal model is:

Q = {born, matured, dead}
Σ = {mature_now, reproduce_now, die_now, north_force, south_force, west_force, east_force, accident}
Γ = {"agent matures", "agent reproduces", "agent dies naturally", "mature agent dies by accident", "new born agent dies by accident", "north_force", "south_force", "west_force", "east_force"}
M = (X, Y), X, $Y \in N_0$, where N_0 is the set of natural numbers including zero.

The initial memory represents the Cartesian coordinates, which describe the position of a cell in a two-dimensional plane (e.g., m_0 = (16, 29)).

q_0 = born
Φ = {grows(mature_now , (X, Y)) = ("agent matures", (X, Y)),
    reproduces(reproduce_now, (X, Y)) = ("agent reproduces", (X, Y)),
    dies(die_now, (X, Y)) = ("agent dies naturally", (X, Y)),
    dies(accident, (X, Y)) = ("mature agent dies by accident", (X, Y)),
    dies_suddenly(accident, (X, Y)) = ("new born agent dies by accident", (X, Y)),
    move_east(west_force , (X, Y)) = ("west_force", (X+1, Y)),
    move_west(east_force , (X, Y)) = ("east_force", (X-1, Y)),
    move_south (north_force , (X, Y)) = ("north_force", (X, Y-1)),
    move_north(south_force , (X, Y)) = ("south_force", (X, Y+1))
    }

Figure 3. The next state transition function of a cell model

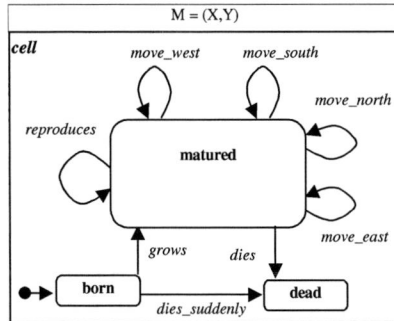

Starting from the initial state q_0 (i.e., *born*) with the initial memory m_0 (i.e., *(16, 29)*), an input symbol $\sigma \in \Sigma$ (i.e., *mature_now*) triggers a function $\varphi \in \Phi$ (i.e., *grows*), which in turn causes a transition (from the set $F(q_0, \varphi)$) to a new state $q \in Q$ (i.e., *matured*) and a new memory value $m \in M$ (in this case, the same *(16, 29)*). The sequence of transitions caused by the stream of input symbols is called a computation. The computation halts when all input symbols are consumed. The result of a computation is the sequence of outputs produced by the sequence of transitions.

Communicating X-Machines

A communicating X-machine model consists of several X-machines that are able to exchange messages. These messages are normally viewed as inputs to some functions of an X-machine model, which in turn may affect the memory structure.

In principle, a communicating X-machine *system* can generally be defined as a tuple $((C_i)_{i=1,\dots,n}, R)$, where C_i is the ith X-machine component that participates in the system, and R is a communication relation between the n X-machines. There are several approaches to formally defining a communicating X-machine. Some of them (Balanescu et al., 1999; Georgescu & Vertan, 2000) deviate from the original definition of the X-machine by introducing communicating functions as well as communicating states. As a result, developing a communicating component is perceived as a modelling activity different than that of developing a stand-alone X-machine. Consequently, one should start from scratch to specify a new component as part of the large system. In addition, components cannot be re-used as stand-alone X-machines or as components of other systems. Moreover, the semantics of the communicating functions impose a limited asynchronous operation of a communicating X-machine.

Definition. In the current chapter, we define a *communicating X-machine system Z* as a tuple, $Z=((C_1,...,C_i,...,C_n), CR)$, where:

- C_i is the ith X-machine component that participates in the system, and
- *CR* is a communication relation between the X-machine components, $CR \subseteq C \times C$, where $C = \{C_1, ..., C_n\}$. *CR* determines the communication channels that exist between the X-machines of the system. A tuple $(C_i, C_k) \in CR$ denotes that X-machine C_i can write a message through a communication channel to a corresponding input stream of X-machine C_k for any $i, k \in \{1, ..., n\}, i \neq k$.

Let *alpha* be a function that returns the alphabet of an input or output stream of an X-machine. If is_i and os_i are the inputs and output streams of an X-machine X_i, then $alpha(is_i) = \Sigma_i$ and $alpha(os_i) = \Gamma_i$.

Definition. The *communicating input streams* structure IS_i of an X-machine X_i is the a *n*-tuple $(is_{i1}, is_{i2}, ..., is_{ij}, ..., is_{in})$ that represents *n* input streams from which the machine X_i can receive messages sent by the other *n-1* X-machines, where:

- is_{ii} is the standard input stream of X_i.
- $is_{ij} = \varepsilon$, if no communication is required with the jth X-machine $(alpha(\varepsilon) = \varnothing)$, or $alpha(is_{ij}) \subseteq \Sigma_i$.

Definition. A *communicating function-input stream mapping* $\Phi IS_i \subseteq \Phi_i \times IS_i$ of an X-machine X_i is a relation between the functions of X_i and the streams from which they accept input.

Definition. The *communicating output streams* structure OS_i of an X-machine X_i is a *n*-tuple $(os_{i1}, os_{i2}, ..., os_{ij}, ..., os_{in})$ that represents *n* output streams to which the machine can send messages to be received by the other *n-1* X-machines, where:

- os_{ii} is the standard output stream of X_i, and

- $os_{ik} = \varepsilon$, if no communication is required with the kth X-machine $(alpha(\varepsilon) = \varnothing)$, or $alpha(os_{ik}) \subseteq \Sigma_k$.

Definition. A communicating function-output stream mapping $\Phi OS_i \subseteq \Phi_i \times OS_i$ of an X-machine X_i is a relation between the functions of X_i and the streams to which they send their output.

Definition. A *communicating X-machine component* C_i is a 10-tuple:

$$C_i = (\Sigma_i,\ \Gamma_i,\ Q_i,\ M_i, \Phi C_i,\ F_i,\ q_{0i},\ m_{0i},\ IS_i,\ OS_i)$$

where ΦC_i is the new set of partial functions that read from either standard input or any other input stream and write to either the standard output or any other output stream.

More specifically, the set ΦC_i consists of four different sets of functions. The notation $(\sigma)_j$ used below denotes an incoming input from X-machine C_j while $(\gamma)_k$ denotes an outgoing output to X-machine C_k. The four different sets of functions are the following:

- The functions that read from standard input stream and write to standard output stream:

  $\varphi_i(\sigma, m) = (\gamma, m')$
  where $\sigma \in \Sigma_i$, $\gamma \in \Gamma_i$, $m,m' \in M_i$, $(\varphi_i \rightarrow is_{ii}) \in \Phi IS_i$ and $(\varphi_i \rightarrow os_{ii}) \in \Phi OS_i$.

- The functions that read from a communication input stream a message that is sent by another X-machine C_j and write to standard output stream:

  $\varphi_i((\sigma)_j, m) = (\gamma, m')$
  where $\sigma \in alpha(is_{ij})$, $\gamma \in \Gamma_i$, $m,m' \in M_i$, $(\varphi_i \rightarrow is_{ij}) \in \Phi IS_i$, $(\varphi_i \rightarrow os_{ii}) \in \Phi OS_i$, and $i \neq j$.

- The functions that read from standard input stream and write to a communication output stream a message that is sent to another X-machine C_k:

  $\varphi_i(\sigma, m) = ((\gamma)_k, m')$

where $\sigma \in \Sigma_i$, $\gamma \in alpha(os_{ik})$, $m,m' \in M_i$, $(\varphi_i \rightarrow is_{ii}) \in \Phi IS_i$, $(\varphi_i \rightarrow os_{ik}) \in \Phi OS_i$ and $i \neq k$.

- The functions that read from a communication input stream a message that is sent by another X-machine C_i and write to a communication output stream a message that is sent to another X-machine C_k:

$$\varphi_i((\sigma)_j, m) = ((\gamma)_k, m')$$
where $\sigma \in alpha(is_{ij})$, $\gamma \in alpha(os_{ik})$, $m,m' \in M_i$, $(\varphi_i \rightarrow is_{ij}) \in \Phi IS_i$, $(\varphi_i \rightarrow os_{ik}) \in \Phi OS_i$ and $i \neq j$, $i \neq k$.

Instead of communicating with only one X-machine C_k, C_i may communicate with C_{k1}, ..., C_{kp}, sending them σ_1, ..., σ_p, respectively; in this case all (C_i, C_j), $1 \leq j \leq p$, must belong to CR and $(\gamma)_k$ will be replaced by $(\gamma_1)_{k1}$ & ... & $(\gamma_p)_{kp}$.

Graphically, if a *solid circle* appears on a transition function, the function accepts input from a communicating stream instead of the standard input stream. It is read as "*reads from*". If a *solid diamond* appears on a transition function, this function may write to a communicating input stream of another X-machine. It is read as "*writes to*" (Figure 4).

Definition. A *communication interface* of an X-machine X_i is the tuple $(IS_i, OS_i, \Phi IS_i, \Phi OS_i)$ where:

- IS_i is the communicating input stream
- OS_i is the communicating output stream
- ΦIS_{ii} is the communicating function-input stream mapping
- ΦOS_i is the communicating function-output stream mapping

Definition. Let X be the set of X-machines, CI the set of communication interfaces ($CI: IS \times OS \times \Phi IS \times \Phi OS$) and C the set of communicating X-machine components. The *communication operator* is defined as:

Communication $COM: X \times CI \rightarrow C$

When the operator is applied,

$$COM ((\Sigma_i, \Gamma_i, Q_i, M_i, \Phi_i, F_i, q_{0i}, m_{0i}), (IS_i, OS_i, \Phi IS_i, \Phi OS_i)),$$

Figure 4. An abstract example of a communicating X-machine component with input and output streams and functions that receive input and produce output in any possible combination of sources and destinations

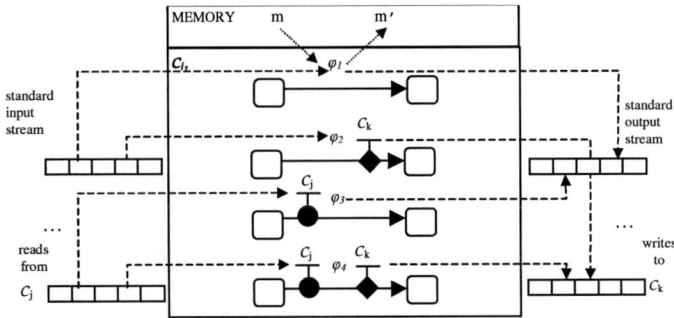

it returns a communicating X-machine component C_i as a tuple:

$$C_i = (\Sigma_i, \Gamma_i, Q_i, M_i, \Phi C_i, F_i, q_{0i}, m_{0i}, IS_i, OS_i).$$

A communicating component C_i contains in its tuple all the necessary compound constructs that integrate an X-machine and its communication interface $(IS_i, OS_i, \Phi IS_i, \Phi OS_i)$. These compound structures are made in such a way that separation to the substructures of which they are made is possible. Therefore, it is feasible to apply the communication operator in reverse, that is, given a communicating component the reverse operation produces the X-machine 8-tuple and its communication interface:

Reverse Communication $COM^{-1}: C \rightarrow X \times CI$

Hereafter, the operator COM will be used with an infix notation (i.e., x COM y instead of $COM (x,y)$) and the operator COM^{-1} will be used with a prefix notation (i.e., $COM^{-1}x$ instead of $COM^{-1}(x)$).

Adding a Biological Clock to a Cell Agent

Time is an important issue in modelling and simulating biology-inspired systems. In cell systems, time plays the role of determining the life state of a cell (Walker et al., 2004). In P systems, computation is based in macro- and microcycles, which drive the evolution of individual membrane regions through the application of rewrite rules (Păun, 2002). Given the X-machine formalism, time can be trivially modelled in the memory structure and altered by clock inputs. New time values could then trigger functions that in turn change the state of the system. This solution assumes that an *advance_time* function will be attached to any state of the machine and a clock tick will be part of the input set; however, such an approach does not contribute to overall abstraction and simplicity. The X-machine model becomes rather complicated and includes functions that deal with time as well as functions that deal with the actual computation, such as the growth of a cell. We suggest that time can be handled separately by modelling a biological clock as an X-machine that communicates with the actual model of the cell.

In the previous example of the cell X-machine, the inputs that trigger the transitions are abstracted events that cause maturity, reproduction, and death of a cell, in a form of explicit input symbols. However, the appropriate time to start the biological processes may be determined by the agent's biological clock, which can be modelled separately. For example, the biological clock is an X-machine of the form:

$Q = \{ticking\};$
$\Sigma = \{tick\};$
$\Gamma = \{mature\_now, reproduce\_now, die\_now\} \cup N_0$
$M = (Clock, MatureAge, ReproductionAge, DieAge)$, where *MatureAge, ReproductionAge, DieAge* $\in N_0$;
$m_0 = (0,15, 40, 100)$, for example
$q_0 = ticking;$
$\Phi = \{$
advance_time(tick, (Clock, MatureAge, ReproductionAge, DieAge)) =
       (Clock+1, (Clock+1, MatureAge, ReproductionAge, DieAge))
               if Clock+1≠ *MatureAge* ∧ *Clock+1* ≠ *ReproductionAge* ∧
               Clock+1≠ *DieAge,*
time_to_grow(tick, (Clock, MatureAge, ReproductionAge, DieAge)) =
       ("mature_now", (Clock+1, MatureAge, ReproductionAge,
       DieAge))
               if Clock +1 = MatureAge,
time_to_reproduce(tick, (Clock, MatureAge, ReproductionAge, DieAge))
=
       ("reproduce_now", (Clock+1, MatureAge, ReproductionAge,
       DieAge))
               if Clock+1 = ReproductionAge,
time_to_die(tick, (Clock, MatureAge, ReproductionAge, DieAge)) =

$$(\text{``die\_now''}, (Clock+1, MatureAge, ReproductionAge, DieAge))$$
$$\text{if } Clock+1 = DieAge$$
}

The next state partial functions F of the biological clock (bc) and the X-machine previously described ($cell$) are shown in Figure 5. The two machines form a communicating system $((C_{cell}, C_{bc}), \{(C_{bc}, C_{cell})\})$. We call this communicating system an X^c-machine (a clocked X-machine). The components of the X^c-machine are constructed through the application of the communication operator COM on bc and $cell$ together with their respective communication interfaces as follows:

$C_{cell} = cell\ COM\ ((\Sigma^*_{cell}, is_{bc}), (\Gamma^*_{cell}, \varepsilon), \{grows \rightarrow is_{bc}, dies \rightarrow is_{bc}, reproduces \rightarrow is_{bc}\}, \{\})$

$C_{bc} = bc\ COM\ ((\varepsilon, \Sigma^*_{bc}), (os_{cell}, \Gamma^*_{bc}), \{\ \}, \{time\_to\_grow \rightarrow os_{cell}, time\_to\_die \rightarrow os_{cell}, time\_to\_reproduce \rightarrow os_{cell}\})$

This application will affect the corresponding functions of the individual components; for example $time\_to\_grow$ and $grows$ become:

$time\_to\_grow(tick, (Clock, MatureAge, ReproduceAge, DieAge)) =$
$(\text{``mature\_now''}_{cell}, (Clock+1, MatureAge, ReproduceAge, DieAge))$

Figure 5. An X^c-machine consisting of the cell agent model and its biological clock

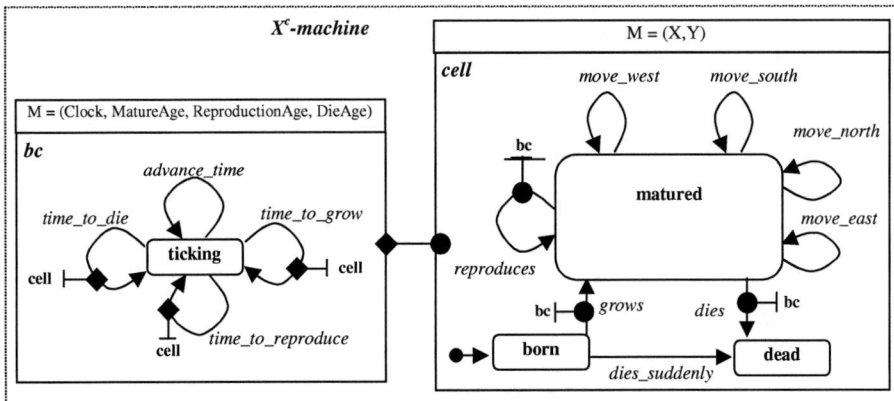

$$if\ Clock\ +1\ =\ MatureAge$$
$$grows(mature\_now_{bc},\ (X,Y)\ )\ =$$
$$(\text{``agent matures''},\ (X,Y)\ )$$

A Tissue Consisting of Individual Cell Agents

So far, we have defined the part of the agent that deals with the agent life state depending on its age. A multi-agent system may be defined in the same way as a communicating X-machine system:

$$Z = ((\ldots,\ agent_i,\ \ldots,\ agent_j,\ \ldots),\ \{\ldots,\ (agent_i,\ agent_j),\ldots\})$$

where the first element is the tuple of the X^c-machines that participate in the system and the second element is the communication relation (Kefalas et al., 2003b). For example, in Figure 1, initially the system is:

$Tissue = ((C_1,\ C_2,\ C_3,\ C_4,\ C_5),\ \{(C_1,\ C_2),\ (C_2,\ C_1),\ (C_1,\ C_3),\ (C_3,\ C_1),\ (C_1,\ C_4),$ $(C_4,\ C_1)\})$

and it is shown as a communicating X-machine in Figure 6 (some states, the biological clocks, and some functions are omitted for the sake of exposition). Communication is added to the *move* functions, as in the example C_1, in which the function *move_south* becomes:

$$move\_south(north\_force_{c2},\ (X,\ Y)) = (\text{``north\_force''}_{c4},\ (X,\ Y-1))$$

which means that *move_south* reads a message from C_2 (*north_force*) and passes its output *north_force* to C_4. This message will imply a change in the position of C_1 and accordingly a change in the position of C_4.

X-machines, as all other formal methods, provide a general framework based on a mathematical formalism that is able to describe what is intended. However, the resulting specification may still be based on over-simplifying assumptions. For example, in the above model, the space is a two-dimensional one; the forces that move the cells can only come from four directions, move steps are discrete as imposed by the two-dimensional Cartesian space, etc. A more detailed model then this is possible but requires extra effort and may result in a more complex model. For example, dealing with simultaneous forces that come from two different directions requires the definition of new functions, which can deal explicitly with this situation. As we have already noted, the model deals implicitly with this situation without deadlocking since the forces

Figure 6. The tissue as a communicating X-machine system

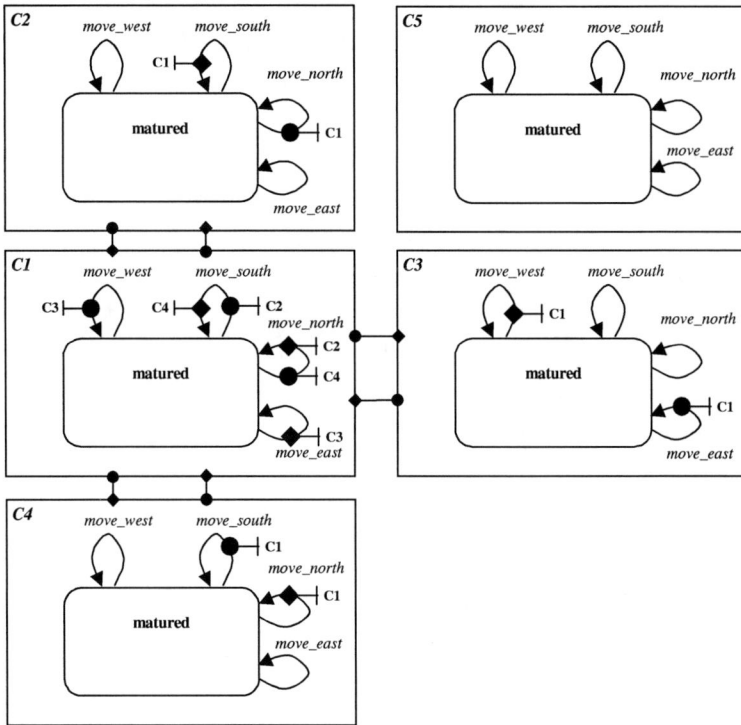

will impose movement of two cells in the same position at a particular instance. But in the long-term, while the system evolves (all inputs are consumed), the system will reach equilibrium and perform as intended. In addition to making the relevant assumptions and abstractions, the resulting specification or model may still contain errors, ambiguities, or inconsistencies. These can be identified through techniques that are used for certifying model correctness as the ones described in the next section.

Modelling and Implementing a Correct System

The previous section demonstrated the usability and flexibility of communicating X-machines as a modelling technique that incrementally develops a complex and complete model of a system. One may argue that the models are still too abstract while the assumptions made are over-simplified. This is true to the extent that someone would like to construct a realistic system for simulation. However, we do not attempt this kind of simulation — something that could be extremely laborious as a task — at present, but instead we try to demonstrate that the method is appropriate to model biology-inspired multi-agent systems. The resulting system would not need to be a simulation of the actual biological process (e.g., simulation of a colony of ants, growth of a cell tissue, etc.), but a system or a technique that could be inspired by the biological process to solve various problems (e.g., ant colony optimization technique, artificial life agents, etc.).

A significant advantage gained by an X-machine model is the ability to use formal verification and testing techniques. Verification and testing support each other and provide confidence in the correctness of the assumptions made. A formal verification technique for X-machines (model checking) is available, which determines if desired properties of a system expressed in a variation of temporal logic, namely *XmCTL*, are satisfied by an X-machine model (Eleftherakis, 2003). Although it is not possible yet to formally verify the complete communicating model, it is possible to apply the model checking technique to the component X-machines (e.g., the biological clock *bc*). Thus, it is possible to verify all the components of a communicating system. For example, one property that someone might want to check against this model is that in all states the property *MatureAge<ReproductionAge<DieAge* is satisfied. This requirement can be expressed as an *XmCTL* formula as:

$$AGMx \ (M(2){<}M(3) \ \wedge \ M(3){<}M(4)),$$

which expresses that all memory instances (operator *Mx*) of all states (operators *A* and *G*) of the X-machine model satisfy that *MatureAge* is less than *ReproductionAge* which is less than *DieAge*. These are represented in the memory tuple in positions 2, 3, and 4, respectively (*M(2),M(3) and M(4)*).

There is a need for formal verification of communicating X-machine models as a whole, and research is under way to extend the existing verification techniques also for communicating X-machine models. With a formal verification technique for the communicating model we will be able to express properties such as "it is impossible for two cells to occupy the same position at the same time".

Having ensured that the model is "correct" with respect to the desired properties, we need to also ensure that the implementation is "correct", this time with respect to the model. This check can be achieved through testing, but only under the important assumption that testing is complete. To guarantee "correctness" of the implementation, one must be certain that all tests are performed, and the results correspond to what the model has specified. As stated before, another strength of using stream X-machines to specify a system is that, under certain well-defined conditions, it is possible to produce a test set that is guaranteed to determine the correctness of an implementation (Ipate & Holcombe, 1997; 1998). The test set produced is proved to find all faults in the implementation. The testing process can therefore be performed automatically by checking whether the output sequences produced by the implementation are identical to the ones expected from the model.

DYNAMIC MODELLING

In the following section, we are going to present a framework that is able to model the part of the system that is responsible for dynamically changing the configuration of the overall multi-agent system Z.

The kind of biology-inspired, agent-based systems we would like to model and eventually simulate are highly dynamic, in the sense that new agents are created, some others cease to exist, and communication channels change over time. It is therefore implied that in such dynamic systems the structure Z, which represents the communicating system, changes. A simplistic approach to tackling this problem is to allow a predefined number of models, arranged in a two- or three-dimensional space, to communicate with their neighbours and exchange messages that, as a result, turn their functionality on and off. This happens, for example, in cellular automata, with the most prominent example being Conway's game of life and its variants (Gardner, 1970).

A more realistic approach would be to model agents (e.g., cells), which are part of biological systems, that are able to pass through a number of life states related to their age, such as being able to grow, reproduce, and die, and at the same time to model the dynamic configuration of the multi-agent system (e.g., tissue), so that each system state has exactly the number of agents and communication channels that represent that particular state.

It is demonstrated in Kefalas et al. (2003c) how modelling of multi-agent systems can be done using the communicating X-machine approach. It is not shown, however, how the communicating X-machine system can alter its configuration when new agents enter or leave the system. The same kind of problem appeared in the attempt for simulating P systems using X-machines (Kefalas et al., 2003a), when a membrane structure alters its configuration by dissolving or dividing its component membranes. The solution provided is that a final "dissolved" state in each corresponding X-machine component and an explicit redirection of communicated objects through the dissolved component.

This solution is, however, not very satisfactory. For the same reasons, membrane division was not addressed.

Summarizing the above issues, in order to model a dynamic biology-inspired system, the X-machine formalism should be able to handle effectively and neatly the following issues:

- Creation of new agent models that participate in the system.
- Deletion of agent models that cease to exist.
- Redirection of communication due to above reasons, but also due to other attributes of the system (e.g., physical movement through space that implies attachment or detachment of agents).

Since new agents are created while others are deleted from the system (e.g., cells are born while others die), X-machine models should be able to specify when and how this could take place. Instead of altering the definition of the X-machine, we suggest that this creation and deletion is handled separately from the agent model. Thus, the control that deals with the reconfiguration of the communicating system can be defined at a meta-level that may include special operators, capable of reconfiguring the system.

Changing a Communicating X-Machine System

In this section, we present four operators that can change the configuration of a communicating X-machine system. While we define only the first operator, we provide examples of applying all of these operators to the cell system previously described. The examples show how operators are applied step-by-step with specific parameters in order to produce a new communicating system.

There are four different operators that deal with reconfiguration of the communicating system: attachment ATT, detachment DET, generation GEN, and destruction DES. In the four definitions that follow P stands for powerset, X is the set of X-machines, C is the set of communicating X-machine components, ΦIS is the set of function-input stream mappings, ΦOS is the set of function-output stream mappings, and Z the set of communication X-machine systems.

Definition. A communicating component may join a set of other components as a result of applying the operator *Attachment* defined as follows:

$$\text{Attachment } ATT: (C \times \Phi IS \times \Phi OS) \times P (C \times \Phi IS \times \Phi OS) \times Z \rightarrow Z$$

A communicating X-machine component C_i joins a set of other components $\{C_j \mid 1 \le j \le n \wedge i \ne j\}$. All components are associated with the functions that should read from or write to the corresponding input and output streams of other components. The result is a new set of communicating components that have the appropriate communication channels between them.

The formal definition of the attachment operator is the following:

$$\forall (c_a, n\varphi is_a, n\varphi os_a): (C, \Phi IS, \Phi OS); \ \forall sc : P (C, \Phi IS, \Phi OS); \ \forall (cx, cr): Z \bullet$$
$$ATT ((c_a, n\varphi is_a, n\varphi os_a), sc, (cx, cr)) = (cx', cr')$$

where:

$$cx' = (cx \setminus (\{c_a\} \cup \{c_b \mid (c_b, n\varphi is_b, n\varphi os_b) \in sc\})) \cup \{c_a'\} \cup nc$$
$$cr' = cr \cup \{(c_a, c_b) \mid (\varphi, is_a) \in n\varphi is_b\} \cup \{(c_b, c_a) \mid (\varphi, is_b) \in n\varphi is_a\}$$

and:

$$COM^{-1} c_a = (x_a, (IS_a, OS_a, \varphi is_a, \varphi os_a))$$
$$c_a' = x_a \ COM \ (IS_a \otimes_{t1} T_{1a}, OS_a \otimes_{t2} T_{2a}, \varphi is_a \cup n\varphi is_a, \varphi os_a \cup n\varphi os_a)$$
$$nc = \{ \ c_b' \mid$$
$$\quad COM^{-1} c_b = (x_b, (IS_b, OS_b, \varphi is_b, \varphi os_b)) \wedge$$
$$\quad c_b' = x_b \ COM \ (IS_b \otimes_{<a>} T_{1b}, OS_b \otimes_{<a>} T_{1b}, \varphi is_b \cup n\varphi is_b, \varphi os_b$$
$$\quad \cup n\varphi os_b)$$
$$\wedge$$
$$\quad (c_b, n\varphi is_b, n\varphi os_b) \in sc\}$$

In this definition, the ordered sequences t_1, t_2, T_{1a}, T_{2a}, T_{1b}, and T_{2b} are defined as:

$$t_1 = ordered \ <i \mid is_i \in ran(n\varphi is_a)>$$
$$t_2 = ordered \ <i \mid os_i \in ran(n\varphi os_a)>$$
$$T_{1a} = ordered \ <is_i \mid is_i \in ran(n\varphi is_a)>$$
$$T_{2a} = ordered \ <os_i \mid os_i \in ran(n\varphi os_a)>$$
$$T_{1b} = ordered \ <is_i \mid is_i \in ran(n\varphi is_b)>$$
$$T_{2b} = ordered \ <os_i \mid os_i \in ran(n\varphi os_b)>$$

where *ordered* is a function that returns the ordered set (sequence) and *ran* denotes the range of a relation set. The operator $\otimes_{<...,i,...>}$ overrides the ith element of the tuple, as in the example $(a, b, c) \otimes_{<1,3>} <z, y> = (z, b, y)$.

Consider Figure 1, where C_5 joins the cluster of cells to the left (C_5 is attached to C_2 and C_3). The operation is defined as:

$ATT\ ((C_5,\ N\Phi IS_5,\ N\Phi OS_5),\ \{(C_2,\ N\Phi IS_2,\ N\Phi OS_2),$
$(C_3,\ N\Phi IS_3,\ N\Phi OS_3)\},\ Tissue) = Tissue'$

where:

$N\Phi IS_5 = \{(move\_east,\ is_2),\ (move\_north,\ is_3)\}$
$N\Phi OS_5 = \{(move\_west,\ os_2),\ (move\_south,\ os_3)\}$
$N\Phi IS_2 = \{(move\_west,\ is_5)\}$
$N\Phi IS_3 = \{(move\_south,\ is_5)\}$
$N\Phi OS_2 = \{(move\_east,\ os_5)\}$
$N\Phi OS_3 = \{(move\_north,\ os_5)\}$

Algorithmically, one can go through the steps of applying the operator one by one as follows:

1. $COM^{-1}\ C_5 = (X_5,\ (IS_5,\ OS_5, \Phi IS_5, \Phi OS_5))$
   Destruct the communicating component C_5 into its X-machine X_5 and its communication interface.
2. $C_5' = X_5\ COM\ (IS_5 \otimes_{<2,3>} <is_2,is_3>,\ OS_5 \otimes_{<2,3>} <os_2,os_3>,\ \Phi IS_5 \cup N\Phi IS_5,\ \Phi OS_5 \cup N\Phi OS_5$
   Construct the new communicating component C_5' by using the X_5 machine together with its new communication interface.
3. $COM^{-1}\ C_2 = (X_2,\ (IS_2,\ OS_2, \Phi IS_2, \Phi OS_2))$
   Destruct the communicating component C_2 into its X-machine X_2 and its communication interface.
4. $C_2' = X_2\ COM\ (IS_2 \otimes_{<5>} <is_5>,\ OS_2 \otimes_{<5>} <os_5>,\ \Phi IS_2 \cup N\Phi IS_2,\ \Phi OS_2 \cup N\Phi OS_2)$
   Construct the new communicating component C_2' by using the X_2 machine together with its new communication interface.
5. $COM^{-1}\ C_3 = (X_3,\ (IS_3,\ OS_3,\ \Phi IS_3,\ \Phi OS_3))$
   Destruct the communicating component C_3 into its X-machine X_3 and its communication interface.
6. $C_3' = X_3\ COM\ (IS_3 \otimes_{<5>} <is_5>,\ OS_3 \otimes_{<5>} <os_5>, \Phi IS_3 \cup N\Phi IS_3, \Phi OS_3 \cup N\Phi OS_3)$
   Construct the new communicating component C_3' by using the X_3 machine together with its new communication interface.
7. $Tissue' = ((C_1, C_2', C_3', C_4, C_5'),\ \{(C_1, C_2'),\ (C_2', C_1),\ (C_1, C_3'),\ (C_3', C_1),\ (C_1, C_4),\ (C_4, C_1),\ (C_2', C_5'),\ (C_5', C_2'),\ (C_3', C_5'),\ (C_5', C_3')\})$
   Construct the new tissue by changing the old communicating components $C_2, C_3,$ and C_5 with the newly constructed ones $C_2', C_3',$ and C_5', and add

the new tuples $(C_2', C_5'), (C_5', C_2'), (C_3', C_5'),$ and (C_5', C_3') in the communication relation.

Definition. A communicating component may depart from a set of other components as a result of applying the operator *Detachment* defined as follows:

Detachment *DET:* $C \times P\ C \times Z \rightarrow Z$

A communicating X-machine component C_i is detached from a set of other components $\{C_j \mid 1 \le j \le n \wedge i \ne j\}$. The result is a new set of communicating components having the appropriate communication channels between them (i.e., the relevant functions that used to read from or write to the detached component are removed). The latter applies to the detached component as well.

For example, consider Figure 1, where C_4 leaves the cluster of cells (C_4 is detached from C_1). The operation is defined as:

$DET(\ C_4,\ \{C_1\},\ Tissue)\ =\ Tissue'$

Algorithmically, the step-by-step application of the operator in this expression is as follows:

1. $COM^{-1}\ C_4 = (X_4, (IS_4,\ OS_4,\ \Phi IS_4,\ \Phi\ OS_4))$
2. $C_4' = X_4 COM(IS_4 \otimes_1 \varepsilon,\ OS_4 \otimes_1 \varepsilon,\ \Phi IS_4 \backslash \{\varphi_i \rightarrow is_1\}, \Phi OS_4 \backslash \{(\varphi_i \rightarrow os_1)\})$
3. $COM^{-1}\ C_1 = (X_1, (IS_1,\ OS_1,\ \Phi IS_1,\ \Phi OS_1))$
4. $C_1' = X_1 COM\ (IS_1 \otimes_4 \varepsilon,\ OS_1 \otimes_4\ \varepsilon,\ \Phi IS_1 \backslash \{\varphi_i \rightarrow\ is_4\}, \Phi OS_1 \backslash \{\varphi_i \rightarrow os_4\})$
5. $Tissue' = ((C_1',\ C_2,\ C_3,\ C_4',\ C_5),\ \{(C_1',\ C_2),\ (C_2,\ C_1'),\ (C_1',\ C_3),\ (C_3, C_1'),\ (C_2,\ C_5),\ (C_5,\ C_2),\ (C_3,\ C_5),\ (C_5,\ C_3)\})$
   where $\backslash$ denotes the set difference.

Definition. A newly constructed X-machine model may join a communicating system as a result of applying the operator *Generation* defined as follows:

Generation *GEN:* $(X \times \Phi IS \times \Phi OS) \times P\ (C \times \Phi IS \times \Phi OS) \times Z \rightarrow Z$

A new communicating X-machine component C_i is created from X, which acts as a genetic code for the new component and joins a set of other components $\{C_j \mid 1 \le j \le n \wedge i \ne j\}$. All components, new and old, are associated with the functions that should read from or write to the corresponding input and output streams of other components. The result is a new set of communicating components having the appropriate communication channels between them. The rest of the components that are not affected are informed about the possibility of having a communication channel open with the new component, by having a new element added to their communicating input and output streams.

Consider Figure 1, where C_6 is born and joins the cluster of cells (C_6 is attached to C_1 and C_4). The operation is defined as:

$GEN ((X_6, N\Phi IS_6, N\Phi OS_6), \{(C_1, N\Phi IS_1, N\Phi OS_1), (C_4, N\Phi IS_4, N\Phi OS_4)\},$
$Tissue) = Tissue'$

where
$N\Phi IS_6 = \{(move\_south, is_1), (move\_north, is_4)\}$
$N\Phi OS_6 = \{(move\_north, os_1), (move\_south, os_4)\}$
$N\Phi IS_1 = \{(move\_north, is_6)\}$
$N\Phi IS_4 = \{(move\_south, is_6)\}$
$N\Phi OS_1 = \{(move\_south, os_6)\}$
$N\Phi OS_4 = \{(move\_north, os_6)\}$

Algorithmically, the step-by-step application of the operator in the above expression is as follows:

1. $C_6' = X_6\ COM ((is_1, \varepsilon, \varepsilon,\ is_4\ \varepsilon, \Sigma_6^*), (os_1,\ \varepsilon,\ \varepsilon,\ os_4, \varepsilon, \Gamma_6^*), N\Phi IS_6, N\Phi OS_6)$
2. $COM^{-1} C_1 = (X_1, (IS_1, OS_1, \Phi IS_1, \Phi OS_1))$
3. $C_1' = X_1\ COM (IS_1 \oplus <is_6>, OS_1 \oplus <os_6>, \Phi IS_1 \cup N\Phi IS_1, \Phi OS_1 \cup N\Phi OS_1)$
4. $COM^{-1} C_4 = (X_4, (IS_4, OS_4, \Phi IS_4, \Phi OS_4))$
5. $C_4' = X_4\ COM (IS_4 \oplus <is_6>, OS_4 \oplus <os_6>, \Phi IS_4 \cup N\Phi IS_4, \Phi OS_4 \cup N\Phi OS_4)$
6. $COM^{-1} C_i = (X_i, (IS_i, OS_i, \Phi IS_i, \Phi OS_i)), i \neq 1, 4, 6$
7. $C_i' = X_i\ COM (IS_i \oplus <\varepsilon>, OS_i \oplus <\varepsilon>, \Phi IS_i, \Phi OS_i), i \neq 1, 4, 6$
8. $Tissue' = ((C_1', C_2', C_3', C_4', C_5', C_6'), \{(C_1', C_2'), (C_2', C_1'), (C_1', C_3'), (C_3', C_1'), (C_1', C_6'), (C_6', C_1'), (C_2', C_5'), (C_5', C_2'), (C_3', C_5'), (C_5', C_3'), (C_4', C_6'), (C_6', C_4')\})$

The operator $\oplus$ adds elements at the end of a tuple, such as in the example $(a, b, c) \oplus <x, y> = (a, b, c, x, y)$.

Definition. A communicating component may cease to exist in a communicating system as a result of applying the operator *Destruction* defined as follows:

Destruction *DES: C× P C× Z→Z*

An existing communicating X-machine component C_i ceases to exist and is detached from a set of other components $\{C_j \mid 1 \le j \le n \wedge i \ne j\}$. The result is a new set of communicating components having the appropriate communication channels between them (i.e., the relevant functions that used to read from or write to the removed component are removed).

Consider Figure 1 where C_1 dies (C_1 is removed and detached from C_2, C_3, and C_6). The operation is defined as:

DES (C_1, $\{C_2$, C_3, $C_6\}$, Tissue) = Tissue'

Algorithmically, the step-by-step application of the operator in the above expression is as follows:

1. *COM^{-1} C_2 = (X_2, (IS_2, OS_2, ΦIS_2, ΦOS_2))*
2. *C_2' = X_2 COM ($IS_2 \otimes_{<1>}$ <ε >, $OS_2 \otimes_{<1>}$ <ε >, $\Phi IS_2 \backslash \{\varphi_i \to is_1\}$, $\Phi OS_2 \backslash \{\varphi_i \to os_1\}$)*
3. *COM^{-1} C_3 = (X_3, (IS_3, OS_3, ΦIS_3, ΦOS_3))*
4. *C_3' = X_3 COM ($IS_3 \otimes_{<1>}$ <ε >, $OS_3 \otimes_{<1>}$ <ε >, $\Phi IS_3 \backslash \{\varphi_i \to is_1\}$, $\Phi OS_3 \backslash \{\varphi_i \to os_1\}$)*
5. *COM^{-1} C_6 = (X_6, (IS_6, OS_6, ΦIS_6, ΦOS_6))*
6. *C_6' = X_6 COM ($IS_6 \otimes_{<1>}$ <ε >, $OS_6 \otimes_{<1>}$ <ε >, $\Phi IS_6 \backslash \{\varphi_i \to is_1\}$, $\Phi OS_6 \backslash \{\varphi_i \to os_1\}$)*
7. *Tissue' = ($\{C_2'$, C_3', C_4, C_5, $C_6'\}$, $\{(C_2'$, $C_5)$, $(C_5$, $C_2')$, $(C_3'$, $C_5)$, $(C_5$, $C_3')$, $(C_4$, $C_6')$, $(C_6'$, $C_4)\}$)*

As we mentioned, our intention is to create a control at a meta-level, and having these special operations will be able to reconfigure the communicating system. There are two primary ways to achieve that:

- Control is global to the whole communicating system.
- Control is attached locally to each agent.

These two methods will be investigated in the following sections. One can imagine a middle approach (by which control stands between agents that communicate, like a bond), but its analysis does not fall under the scope of this chapter.

Modelling a Dynamic System: The Global Approach

Consider a multi-agent system model created by developing a communicating X-machine Z, using X-machine components C_i that communicate directly using the approach described above. In this first approach, the information that drives the dynamic evolution of the system is global and can be described at a meta-level as:

- A set of meta-rules (i.e., rules that refer to the configuration of the system).
- A meta-machine (i.e., an X-machine-like device that manipulates the components or attributes of the system).

Conceptually, the actual communicating system and the meta-system could be considered to play the role of the environment to each other. For example, the global control receives input from its environment, in this case Z, processes this input, and returns an output in form of operations described above, which in turn change the environment (Figure 7).

The control device should be able to control the change of the configuration of the communicating system, and therefore should possess the following information at any time:

- The tuple that represents the communicating system
  $Z = ((C_1, ..., C_i, ..., C_n), CR)$.
- The current system state (SZ) of Z. SZ is defined as a set of tuples $SZ = \{... (q_c, M_c, \varphi_c)_i, ...\}$ with each tuple corresponding to a component

Figure 7. Global approach for modelling a dynamic system

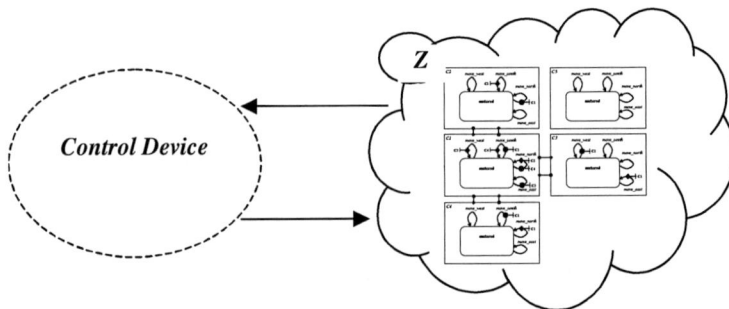

C_i: the current state (q_c) of C_i, the current memory (M_c) of C_i, and the last function (φ_c) that was applied in C_i.

- Definitions of all components that exist or may be added to the system. These definitions act as genetic codes (GC) for the system. GC is a set of tuples, $GC = \{\dots (\Sigma, \Gamma, Q, M, \Phi, F, \Phi_R, \Phi_W)_p \dots\}$ where the first six elements are as defined in the definition of the X-machine given in the previous section, and the last two are the set of functions that may be involved in communication with other components (i.e., Φ_R may read from communicating streams and Φ_W may write to communicating streams).

Using this information, the control device can perform the following actions:

- Generate a new component and attach it to the communicating machine Z.
- Destruct an existing component of Z and rearrange the communication of the rest components appropriately.
- Change the communication between a group of components because of the addition or removal of an existing component to or from the communicating cluster (attachment, detachment).

All of these actions are possible through the use of the operators that were presented in the previous section. For example, in Figure 1, consider just the first case, where the cell C_5 is attached to the group of cells because it receives a force from the east that causes C_5 to move to the west. The move implies that C_5 has to initiate communication with C_2 and C_3, since in this case the criterion for establishing communication is the immediate neighbourhood. The following is a step-by-step description of how this is possible.

The state of the communicating system SZ is updated because of the movement of C_5. The control device identifies the attachment of this component to the other two by comparing the coordinates of this cell with the coordinates of the rest of the cells (all represented as components of Z). This identification is possible because all the required information is available through the memory instances of all of the components in SZ. Comparing the coordinates, the control device identifies that C_5 is now next to C_2 and C_3, and therefore it should apply the appropriate operators that change Z to a new, evolved, communicating system Z', in this case by just altering the communication relation CR. For this example, the attachment operator should be used as follows:

ATT $((C_5,$ $N\Phi IS_5,$ $N\Phi OS_5),$ $\{(C_2,$ $N\Phi IS_2,$ $N\Phi OS_2),$ $(C_3,$ $N\Phi IS_3,$ $N\Phi OS_3)\},$ $Z)$ = Z'

To construct $N\Phi IS_i$ the genetic code of the component representing the cell C_i should be known, as information is available to the control machine by the set GC. For example, for C_5, the genetic code dictates that the set of the functions that may read from a communicating stream is Φ_{R5} = *{move_east, move_west, move_south, move_north}* and the set of functions that may write to a communicating stream is Φ_{W5} = *{move_east, move_west, move_south, move_north}*. Reasoning about the particular system and taking into account the particular rules of communication, it is now possible to create $N\Phi IS_5$ = *{(move_east$\to$ is_2), (move_north$\to$ is_3)}*, and $N\Phi OS_5$ = *{(move_west$\to$ os_2), (move_south$\to$ os_3)}*. The same reasoning is applied to construct $N\Phi IS_2$, $N\Phi OS_2$, $N\Phi IS_3$, and $N\Phi OS_3$.

Modelling a Dynamic System: The Local Approach

The second approach supports a local view for control, meaning that each component of the communicating X-machine system Z is wrapped around an interface device. This is a specialised communicating device that is able to communicate with all other interfaces. As above, the device is defined at a meta-level for each component either as a set of meta-rules or as a meta-machine.

The interface should be able to control the change of the configuration of the communicating system, and therefore should possess information similar to the global control, but this time having only the local view of the component C_i that this interface controls:

- The tuple that represents part of the communicating system Z_p = $((C_1, ..., C_i, ..., C_n), CR_p)$ (i.e., not all components but only the subset CR_p of CR that refers to the communication of C_i).
- The current component state (SC_i) of C_i. SC_i is defined as a tuple SC_i = $(q_c, M_c, \varphi_c)_i$, with each element of the tuple representing the current state (q_c) of C_i, the current memory (M_c) of C_i, and the last function (φ_c) that was applied in C_i.
- The genetic codes (GC_i) for SC_i. GC_i is a tuple, and GC_i = $(\Sigma, \Gamma, Q, M, \Phi, F, \Phi_R, \Phi_W)_i$, as before.

Communication between components of the system is only possible through their interfaces. Therefore, we have an explicit one-to-all communica-

Figure 8. Local approach for modelling a dynamic system

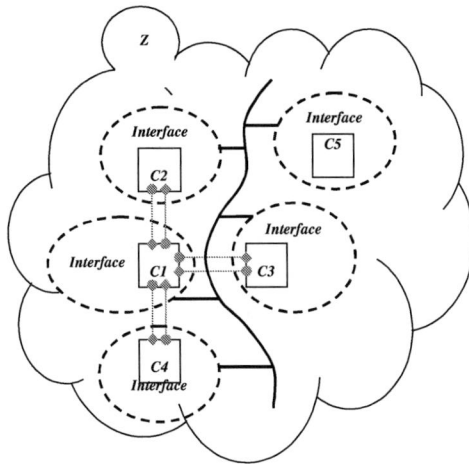

tion between interfaces and implicit one-to-many communication between components (Figure 8). An interface broadcasts control messages that are received by all others, but it can also broadcast messages coming from its local component C_i that are received only by the interfaces whose components are supposed to communicate with C_i. As before, the interface applies operators to the system and changes it. The only difference with the global control is that now information is only locally available, and whenever extra information is needed, it can be retrieved by exchanging control messages with the other interfaces.

DISCUSSION

The approach described in this chapter aims to develop formal models of biology-inspired agents. There are two fundamental concepts associated with any dynamic or reactive system, such as an agent, that is situated in some environment and reacting with it. First, it is the environment itself, which could be precisely or ill-specified or even completely unknown, and involves identifying its important aspects and the way in which they may change in accordance with the activities of the agent. Second, it is the agent, which responds to environmental changes by changing its basic parameters and possibly affecting the environment as well. Thus, there are two ways in which the agent reacts: it undergoes internal changes, and it produces outputs that affect the environment.

The mentioned concepts are captured by X-machines as demonstrated by the example presented. X-machine models can be created as stand-alone agent models. These models can be verified through XmCTL and model checking algorithms and tested, so that their "correctness" is proven. It is then possible to construct a communicating agent model through the use of communicating streams. The important issue addressed is that the communicating components can be derived from stand-alone models when combined with their communication interface. This combination has a tremendous practical advantage in the sense that the tools constructed for X-machines can incrementally develop communicating systems without destroying stand-alone models (Kefalas et al., 2003b). The X-machine method is rather intuitive, while formal descriptions of data types and functions can be expressed in any known mathematical notation. However, X-machine modelling tools support a specific notation, namely XMDL, which provides all the necessary constructs for mathematical modelling through simple text (Kefalas & Kapeti, 2000).

We have also defined a set of operators that are able to reconfigure a communicating agent system. These operators are going to be useful when modelling of the control or interface devices is attempted, either at a global or at a local level. Actually, the operators are actions of the control or interface devices through which the evolution of the communicating system takes place. Defining such devices or interfaces, either in the form of meta-rules or meta-machines, is rather complex and specific to the problem at hand. This is because modelling of such a device in our example falls out of the scope of this chapter. The global and the local approaches have their usual advantages and disadvantages — that is, completeness and availability of information versus communication overhead.

However, bearing in mind that the model should lead to the implementation of such systems, the implementer may choose the most appropriate technique for modelling the control or interface device, or even to combine both approaches into a hybrid one.

CONCLUSION

In this chapter we have showed how modelling of multi-agent systems can be achieved through the use of communicating X-machines and how four meta-operators can change the configuration of the system. By taking as an example a biology-inspired multi-agent system, we demonstrated that control over the global system state should be modeled in a way that does not affect modelling of individual components of the system. The important contribution of this work is that overall modelling of multi-agent systems can be achieved by:

- Decoupling agent and time modelling by using the communicating X-machine approach.
- Decoupling modelling of agents and multi-agent reconfiguration by using operators at a meta-level, which will be responsible for dynamically configuring the system as time progresses.

This process allows a modular approach towards modelling, since the components of a multi-agent system can be modelled separately, as well as a disciplined approach, since the modelling activity can be performed in stages. By having minimal changes in the X-machine model, we are still able to exploit, to the maximum possible extent, the legacy of X-machine theory concerning complete testing and formal verification for each individual component. Such models may comprise a library of off-the-shelf, ready-made, verified, and tested components, which may be re-used in other systems. In the first stage of this process, individual models are developed, and the whole multi-agent system is constructed. From a practical point of view, this creates a significant advantage to the modeler.

Future work involves the development of a theory for verifying and testing communicating systems as a whole. It is also worth investigating how complex it is to model actual biological processes through formal methods and what are the benefits of such modelling. Finally, the tools for X-machines, apart from animation of stand-alone components, need to be equipped with animation of whole communicating systems.

REFERENCES

Adleman, L.M. (1994). Molecular computation of solutions to combinatorial problems. *Science*, 226, 1021-1024.

Attoui, A. & Hasbani, A. (1997). Reactive systems developing by formal specification transformations. *Proceedings of the 8th International Workshop on Database and Expert Systems Applications (DEXA 97)*, 339-344.

Balaneascu, T., Cowling, A.J., Georgescu, H., Gheorghe, M., Holcombe, M., & Vertan, C. (1999). Communicating stream X-machines systems are no more than X-machines. *Journal of Universal Computer Science*, 5 (9), 494-507.

Balanescu, T., Gheorghe, M., Holcombe, M., & Ipate, F. (2002). A variant of EP systems. Pre-proceedings of the *Workshop on Membrane Computing*, Curtea de Arges, Romania, Publication No.1, MolCoNet project – IST – 2001 – 32008, 57-64.

Banatre, J-P. & Le Metayer, D. (1990). The gamma model and its discipline of programming. *Science of Computer Programming*, 15, 55-77.

Benerecetti, M., Giunchiglia, F., & Serafini, L. (1999). A model-checking algorithm for multi-agent systems. In Muller, J. P., M. P. Singh, M. P., & Rao, A. S. (Eds.), *Intelligent Agents V. Lecture Notes in Artificial Intelligence, 1555*, Berlin: Springer, 163-176.

Berry, G. & Boudol, G. (1992). The chemical abstract machine. *Theoretical Computer Science*, 96, 217-248.

Brazier, F., Dunin-Keplicz, B., Jennings, N., & Treur, J. (1995). Formal specification of multi-agent systems: A real-world case. *Proceedings of the International Conference on Multi-Agent Systems (ICMAS'95)*, Cambridge, MA: MIT Press, 25-32.

d'Inverno, M., Kinny, D., Luck, M., & Wooldridge, M. (1998). A formal specification of dMARS. In Singh, M. P., Rao, A., & Wooldridge, M. J. (Eds.), *Intelligent Agents IV*. Lecture notes in *Artificial Intelligence*, 1365, Berlin: Springer, 155-176.

Dorigo, M., Maniezzo, V., & Colorni, A. (1996). The ant system: Optimisation by a colony of co-operating agents. *IEEE Transactions on Systems, Man and Cybernetics*, 26 (1), 1-13.

Duan, Z. H., Holcombe, M., & Linkens, D. A. (1995). Modelling of a soaking pit furnace in hybrid machines. *System Analysis, Modelling, Simulation*, 18, 153-157.

Eilenberg, S. (1974). *Automata, languages and machines, vol. A*. Orlando, FL: Academic Press.

Eleftherakis, G. (2003). *Formal verification of X-machines models: Towards formal development of computer-based systems*. PhD Thesis, University of Sheffield, Department of Computer Science.

Eleftherakis, G. & Kefalas, P. (2001). Model checking safety critical systems specified as X-machines. Analele Universitatii Bucharest, *Mathematics-Informatics series*, 49 (1), 59-70.

Fisher, M. & Wooldridge, M. (1997). On the formal specification and verification of multi-agent systems. *International Journal of Cooperating Information Systems*, 6 (1), 37-65.

Futatsugi, K., Goguen, J., Jouannaud, J.P., & Meseguer, J. (1985). Principles of OBJ2. In Reid, B. (Ed.), *Proceedings of the Twelfth ACM Sympo-*

sium on Principles of Programming Languages, Association for Computing Machinery, 52-66.

Gardner, M. (1970). Mathematical games: The fantastic combinations of John Conway's new solitaire game "life". *Scientific American*, 223, 120-123.

Georgescu, H. & Vertan, C. (2000). A new approach to communicating X-machines systems. *Journal of Universal Computer Science, 6* (5), 490-502.

Gheorghe, M., Holcombe, M., & Kefalas, P. (2001). Computational models of collective foraging. *BioSystems*, 61, 133-141.

Gheorghe, M., Holcombe, M., & Kefalas, P. (2003). Eilenberg P systems: A bio-computational model. *Proceedings of the 1st Balkan Conference in Informatics*, November 2003, 147-160.

Harel, D. (1987). Statecharts: A visual approach to complex systems. *Science of Computer Programming, 8* (3), 231-274

Holcombe, M. (1988). X-machines as a basis for dynamic system specification. *Software Engineering Journal, 3* (2), 69-76.

Holcombe, M. (2001). Computational models of cells and tissues: Machines, agents and fungal infection. Briefings in *Bioinformatics*, 2, 271-278.

Holcombe, M. & Ipate, F. (1998). *Correct systems: Building a business process solution.* London: Springer-Verlag.

Ipate, F. & Holcombe, M. (1997). An integration testing method that is proved to find all faults. *International Journal of Computer Mathematics, 63* (3), 159-178.

Ipate, F. & Holcombe, M. (1998). Specification and testing using generalised machines: A presentation and a case study. *Software Testing, Verification and Reliability*, 8, 61-81.

Jennings, N.R. (2000). On agent-based software engineering. *Artificial Intelligence*, 117, 277-296.

Jones, C. B. (1990). *Systematic software development using VDM* (2nd ed.). Englewood Cliffs, NJ: Prentice-Hall.

Kauffman, S.A. (1993). *The origins of order: Self-organization and selection in evolution.* Oxford: Oxford University Press.

Kefalas, P. (2002). Formal modelling of reactive agents as an aggregation of simple behaviours. In Vlahavas, I. P. & Spyropoulos, C. D. (Eds.), *Lecture Notes in Artificial Intelligence*, 2308, Berlin: Springer, 461-472.

Kefalas, P. & Kapeti, E. (2000). A design language and tool for X-machines specification. In Fotiadis, D. I. & Nikolopoulos, S. D. (Eds.), *Advances in informatics.* Singapore: World Scientific Publishing Company, 134-145.

Kefalas, P., Eleftherakis, G., Holcombe, M., & Gheorghe, M. (2003a). Simulation and verification of P systems through communicating X-machines. *BioSystems*, *70* (2), 135-148.

Kefalas, P., Eleftherakis, G., & Kehris, E. (2003b). Communicating X-machines: A practical approach for formal and modular specification of large systems. *Journal of Information and Software Technology*, *45* (5), 269-280.

Kefalas, P., Holcombe, M., Eleftherakis, G., Gheorghe, M. (2003c). A formal method for the development of agent based systems. In Plekhavona, V. (Ed.), *Intelligent agent software engineering*. Hershey, PA: Idea Group Publishing, 68-98.

Leiden University (2003). *A Bibliography of Molecular Computation and Splicing Systems*. *http://www.wi.leidenuniv.nl/~pier/dna.html*

Odell, J., Parunak, H. V. D., & Bauer, B. (2000). Extending UML for agents. *Proceedings of the Agent-Oriented Information Systems Workshop at the 17th National conference on Artificial Intelligence*, 3-17.

Păun, Gh. (2000). Computing with membranes. *Journal of Computer and System Sciences*, *61* (1), 108-143.

Păun, Gh. (2002). *Membrane computing: An introduction*. Berlin: Springer.

Rao, A. S. & Georgeff, M. P. (1993). A model-theoretic approach to the verification of situated reasoning systems. In Bajcsy, R. (Ed.), *Proceedings of the 13th International Joint Conference on Artificial Intelligence (IJCAI'93)*. San Francisco: Morgan Kaufmann, 318-324.

Rao, A. S., & Georgeff, M. P. (1995). BDI agents: From theory to practice. *Proceedings of the 1st International Conference on Multi-Agent Systems (ICMAS-95)*, 312-319.

Reisig, W. (1985). Petri nets: An introduction. *EATCS Monographs on Theoretical Computer Science*, 4, Berlin: Springer.

Rosenschein, S. R. & Kaebling, L. P. (1995). A situated view of representation and control. *Artificial Intelligence*, *73* (1-2), 149-173.

Spivey, M. (1989). *The Z notation: A reference manual*. Englewood Cliffs, NJ: Prentice-Hall.

Università degli Studi Milano Bicocca. (2003). *P Systems*. *http://psystems.disco.unimib.it*

Walker, D. C., Holcombe, M., Southgate, Mac Neil, S., J. & Smallwood, R. H. (2004). Agent-based modeling of the social behaviour of cells. *BioSystems*. (To appear).

Wulf, W. A., Shaw, M., Hilfinger, P. N., & Flon, L. (1981). *Fundamental structures of computer science*. Boston: Addison-Wesley.

About the Authors

Marian Gheorghe is a senior lecturer in the Department of Computer Science of the University of Sheffield, UK. He holds a BSc in mathematics and computer science and a PhD in computer science both from Bucharest University, Romania. Gheorghe has research interests in computational models, including both theoretical and applicative fields of computer science. His research interests cover generative power of various formal grammars and automata, closure properties of the families of languages generated by them, formal specification and testing, natural computing, systems biology, and formal specification models of agent-based systems. He has published nearly 50 papers in various journals and international conference proceedings. Gheorghe co-edited a special issue of *BioSystems* on relationships between biological modeling and theoretical linguistics, especially formal language theory. He is a member of EMCC, EATCS and is participating in the European network of excellence, Molecular Computing Networking.

****

Gabriel Ciobanu graduated from Iasi University, obtained a PhD from the same university, and now holds a professorship in the Department of Computer Science at Iasi University, Romania. He has a wide range of interests in computing, including distributed systems and computational methods in biology. He is the author of over 80 papers in computer science and mathematics, co-editor of three volumes, author of a book on programming semantics and co-author of a book on network programming. He has received public recognition for his research on several occasions, including the Romanian

Academy's "Grigore Moisil" Award for Computer Science in 2000, a Japan Society for the Promotion of Science Fellowship in 1995, a DAAD Research Fellowship in 1994, and a Royal Society of London Fellowship in 1992. He is a member of the following associations and societies: EATCS, EAPLS, ISCB, EMCC, and AMS.

George Eleftherakis holds a BSc in physics, and an MSc and PhD in computer science from the University of Sheffield, UK. His current interests are in formal methods and especially in formal verification techniques. He is conducting research in the area of temporal logic and model checking and particularly in the development of a formal verification technique for the X-machine formal model. He also has special interest for Internet applications and has dealt with network programming and databases, especially using Java. He has published over 15 papers, especially in the area of formal methods. He is now a lecturer in the Computer Science Department at CITY College, an affiliated institution of the University of Sheffield in Greece.

Andrés Cordón Franco is assistant professor in computer science and artificial intelligence at the University of Sevilla, Spain. He is a member of the Research Group on Natural Computing and also of the European Molecular Computing Consortium. His main research interests are mathematical logic, logic in computer science, and membrane computing, both at the theoretical level and at the level of software simulation. He has actively collaborated with other members of the Research Group on Natural Computing and co-authored several contributions to international journals.

Giuditta Franco graduated with a degree in mathematics from the University of Pisa, Italy, in 2001. Since 2003, she has been a PhD student in the Department of Computer Science at the University of Verona, Italy. The focus of her research is molecular computing, in particular DNA computing and membrane computing. She has attended several international meetings, has published several papers in this field, and is a member of EMCC (the European Molecular Computing Consortium). Her current research focuses on the informational analysis of DNA sequences and the modelling of the dynamical features of immune processes by using membrane systems.

Jean-Louis Giavitto, a computer scientist at CNRS (a French research institute), is the director of the Computer Science Laboratory at the University of Évry Val d'Essonne, France. He is the editor-in-chief of *Technique et*

Science Informatique, a French scientific journal in informatics. After receiving an engineering diploma at ENSEEIHT in programming languages, he worked in the software research group of Alcatel before designing an experimental data-parallel language for modeling and simulating dynamical systems during his PhD studies. He is the editor of two books and several chapters and has published more than 40 research papers. His interests include unconventional programming paradigms (especially inspired by biological systems), space and time representation in computer science, and the use of topological notions in programming languages.

Richard Gregory is currently involved in the reengineering of the COSMIC project with the Biocomputing and Computational Biology Research Group in the Department of Computer Science at the University of Liverpool, UK. He gained his BSc in computer science at the University of Liverpool in 1999. From 1999 to 2004 he completed a PhD in the fields of computer science, biology, and engineering, which led to the ambitious project of microbial evolution in an individual-based architecture. This project later became known as COSMIC.

Mike Holcombe is a professor of computer science and the dean of the faculty of engineering at the University of Sheffield, UK. He has conducted research in software and systems engineering, software testing, formal methods in systems engineering, formal specification and test generation for software and systems, requirements engineering, specification and analysis of hybrid systems, theoretical computer science, algebraic theory of general machines, formal models of user behaviour and human-computer interface design, visual formal specification languages and visual reasoning, biological systems, biocomputing and the computational modeling of cellular processing, metabolic systems theory, computational models of immunological systems, and developmental modeling of plant growth. He has published over 100 papers and six books on theoretical computer science, software engineering, and computational biology.

Lila Kari received her MSc in mathematics and computer science at the University of Bucharest, Romania, and her PhD in mathematics and computer science at the University of Turku, Finland. She is the recipient of numerous research and teaching awards including the Canada Research Chair in Biocomputing for 2001 to 2006. Her current research interests include biomolecular computing, formal languages, automata theory, and the theory of self-assembly.

Petros Kefalas is the vice principal at CITY College, Thessaloniki, Greece. He holds an MSc in artificial intelligence and a PhD in computer science, both with the University of Essex. He conducted research in parallel logic programming and search algorithms in artificial intelligence. He has published over 40 papers in journals and conference proceedings and co-authored a Greek textbook in artificial intelligence. He is currently involved in investigating the applicability of formal methods for specifying, verifying, and testing agent systems. He is currently a reader in the Computer Science Department of CITY College, which is an affiliated institution of the University of Sheffield. He teaches logic programming, artificial intelligence, and intelligent agents.

Elena Losseva received her Bachelor of Mathematics from the University of Waterloo, Canada, in 2000. She has worked on projects at IBM in the compiler optimization and in DB2 teams and is currently pursuing a PhD in computer science at The University of Western Ontario in London, Canada, under the supervision of Professor Lila Kari. Her research interests are in the field of DNA computing and formal languages. Losseva's most recent research includes a study of reliability issues in DNA computing. In 2002, she visited the California Institute of Technology for the Computing Beyond Silicon study program. She also collaborates with IBM's Women in Technology (WIT) project.

Vincenzo Manca graduated in mathematics in 1971 from the University of Pisa, Italy, which he was with until 1988, when he moved to the University of Udine, Italy, as an associate professor. In 2002, he joined the University of Verona, Italy, as full professor with the Computer Science Department. He has tutored more than 30 MSc and PhD students, and his research interests include unconventional computation models (DNA computing, molecular computing) and the informational analysis of biological systems. His latest research endeavour is the coordination of a joint group of computer scientists and biologists running biotechnological experiments on DNA extraction algorithms. He is the author of over 60 international scientific publications.

Carlos Martín-Vide is a professor and head of the Research Group on Mathematical Linguistics at Rovira i Virgili University, Tarragona, Spain. His areas of expertise are formal language theory and mathematical linguistics. His last volumes edited are *Where Mathematics, Computer Science, Linguistics and Biology Meet* (Kluwer, 2001, with V. Mitrana), *Grammars and Automata for String Processing: From Mathematics and Computer Science* *to Biology, and Back* (Taylor and Francis, 2003, with V. Mitrana), and *Formal Languages and Applications* (Springer, Berlin, 2004, with V. Mitrana and G. Păun). He has published more than 210 papers in conference proceedings and peer-reviewed journals. He is the editor-in-chief of the journal *Grammars*, and the chairman of the International PhD School in Formal Languages and Applications.

Olivier Michel is an assistant professor at the University of Évry Val d'Essonne, France. During his PhD studies he has developed several formalisms to handle dynamical space in declarative languages. He has published more than 30 research papers in the areas of compiler construction, data-flow, and declarative languages. His interests include bio-inspired programming languages, tools, and techniques for dynamical systems representation using dynamical structures.

Victor Mitrana is a professor at the Faculty of Mathematics and Computer Science, University of Bucharest, Romania, and a Ramon y Cajal professor at the Research Group on Mathematical Linguistics at Rovira i Virgili University, Tarragona, Spain. He has published more than 150 papers and three books (two in Romanian and one in English), and edited three books. His fields of interest are formal languages and automata (theory and applications), molecular computing, algorithms, and combinatorics on finite/infinite sequences. He is an associate editor for the *Journal of Applied Mathematics and Computing* and editor of the *Journal of Universal Computer Science*. He was a Humboldt Fellow in 1995, 1996, and in 1999 and received the Gheorghe Lazar prize of the Romanian Academy in 1999.

Miguel Angel Gutiérrez Naranjo is an assistant professor in computer science and artificial intelligence at the University of Sevilla, Spain. He holds a PhD in mathematics and is a member of the Research Group on Natural Computing and the European Molecular Computing Consortium. His main interests are machine learning, logic programming, and membrane computing. He focuses his research both on mathematical aspects of these areas and computer implementations. He has co-authored several chapters of scientific books and many papers in international journals.

John W. Palmer has undergraduate degrees in civil engineering from the University of Nottingham, UK, and in physics from the University of Manchester, UK. His PhD thesis, also undertaken at Manchester, was in the field of observational astronomy. At present, he is a research fellow at the University of Liverpool with particular interest in the application of computational grid technology to biocomputing and optimization.

Ray C. Paton is a reader in the Department of Computer Science at the University of Liverpool, UK, where he heads the Biocomputing and Computational Biology Research Group. His research interests include biologically inspired computation, theoretical and computational biology, and knowledge modelling. Paton has been instrumental in the development of a number of multidisciplinary initiatives including the international IPCAT conferences and the U.K. EPSRC-funded CytoCom and MIPNETS networks. He is an editor of *BioSystems* and of a book series titled *Studies in Multidisciplinarity*. Paton is a guest faculty member of the Statistical Sciences Group at Los Alamos National Laboratory.

Gheorghe Păun graduated from the Faculty of Mathematics, University of Bucharest, in 1974 and received his PhD in mathematics from the same university in 1977. Since 1990, he has been a senior researcher at the Institute of Mathematics of the Romanian Academy. He has visited numerous universities in Europe, Asia, and North America, and his main research areas are formal language theory and its applications, computational linguistics, DNA computing, and membrane computing (a research area initiated by him, belonging to natural computing). He has published over 400 research papers (collaborating with many researchers worldwide), has lectured at over 100 universities, and has given numerous invited talks at recognized international conferences. He has published 11 books in mathematics and computer science and has edited more than 25 collective volumes. He is on the editorial boards of 14 international journals in computer science and has been involved in the program and steering committees for many recognized conferences and workshops. In 1997, he was elected a member of the Romanian Academy.

Mario J. Pérez-Jiménez is a professor in computer science and artificial intelligence at the University of Sevilla, Spain, where he is the head of the Group of Research in Natural Computing. He has been responsible for a project supported by the European Community ("Finite Model Theory and Bounded Arithmetic") and is now working on a project supported by Ministerio de Ciencia y Tecnologia of Spain (*"Desarrollo, verificacion y automatizacion de modelos moleculares y celulares con membranas"*). He has published eight books of mathematics and computer science, and more than 80 scientific papers in prestigious scientific journals. He is a member of European Molecular Computing Consortium.

Agustín Riscos-Núñez has held a research fellowship as a doctorate student with the Department of Computer Science and Artificial Intelligence at the University of Sevilla, Spain, since November 2001. He is a member of the Research Group on Natural Computing and the European Molecular Computing Consortium. His main interests are in complexity theory, membrane computing, and other areas of natural computing, both at the theoretical level and the level of computer simulation. In the last year he has actively collaborated with the members of the Research Group, submitted papers to several international conferences, and published in international journals.

Jon R. Saunders received a BSc in microbiology from the University of Bristol, UK in 1970, and a PhD in microbial genetics, also from the University of Bristol, in 1973. After postdoctoral work in the Department of Bacteriology at the University of Bristol, he was appointed a lecturer in microbiology at the University of Liverpool, UK. He has held the chair of microbiology since 1992 and is dean of science at the University of Liverpool. His research interests are in the areas of plasmids, bacteriophages, and other mobile genetic elements in relation to bacterial pathogenesis, biogeochemical cycling, and gene transfer in the environment, both practically and in relation to biocomputing.

Giuseppe Scollo graduated *cum laude* in electrical engineering in 1977 at the University of Catania, Italy, where he worked until 1986, when he moved to the University of Twente, The Netherlands, as an associate professor until 1994. He earned a PhD in 1993 from Twente University and carried out research and scientific consultancy on algebraic methods in software engineering and language processing until 2001, when he joined the University of Verona, Italy, as an associate professor with the Computer Science Department. His research interests include algebraic methods in software design and in language processing, testing theory, and discrete dynamic systems. He is the author of over 60 international scientific publications.

Petr Sosík received his PhD in computer science in 1997 in Prague. Since 1992 he has been teaching as an assistant professor (and later associated professor) at the Silesian University at Opava, Czech Republic. Since 2003 he has been visiting the University of Western Ontario, London, Canada, as a postdoctoral fellow. His research interests include formal language-based models of multi-agent and biological systems, with applications in biocomputing, artificial intelligence, and cognitive science.

Ioanna Stamatopoulou holds a BSc (Hons) degree in computer science from the University of Sheffield, UK, and an MSc in artificial intelligence from the University of Edinburgh, UK. Her current interests are in formal methods and especially in formal modelling of multi-agent systems. She is conducting research in the area of communicating X-machines and molecular computing approaches such as P-systems. She is now a PhD candidate and a tutor in the Department of Computer Science at CITY College, an affiliated institution of the University of Sheffield in Greece.

Costas Vlachos received a BEng in applied electronics from the Technological Educational Institution of Athens, Greece in 1993, and an MSc and PhD in automatic control systems from Liverpool John Moores University, UK in 1995 and 2000, respectively. From 1998 to 2001, he worked with the Control Systems Research Group at Liverpool John Moores University as a research assistant. He is currently with the Biocomputing and Computational Biology Research Group in the Department of Computer Science at the University of Liverpool, UK. His research interests are in the areas of adaptive and nonlinear control systems theory, robust and optimal control, and global optimization methods with an emphasis on evolutionary algorithms.

Q. H. Wu earned an MSc (Eng) in electrical engineering from Huazhong University of Science and Technology (HUST) in China in 1981. From 1981 to 1984, he was appointed lecturer in electrical engineering at this same university. He obtained a PhD degree from The Queen's University of Belfast (QUB), UK, in 1987. He worked as a research fellow and senior research fellow at QUB from 1987 to 1991, and lecturer and senior lecturer in the department of mathematical sciences at Loughborough University, UK from 1991 to 1995. Since 1995 he has held the chair of electrical engineering in the Department of Electrical Engineering and Electronics at the University of Liverpool, UK, acting as the head of the Intelligence Engineering and Automation group. Wu is a chartered engineer, fellow of IEE, and senior member of IEEE. His research interests include adaptive control, neural networks, learning systems, multi-agent systems, computational intelligence, biocomputation, and power systems control and operation.

Index

N

natural computing 1, 2
network of communicating automata 224
NFAT/AP1 240
NP-complete problem 122

O

OBJ 247

P

P systems 3, 247
P systems with input 122
parallel processing 9
periodicity 35
Petri nets 247
population size 211
Prolog simulator 115
PRS 247

Q

quorum sensing 209

R

random-context conditions 84
real-world problems 213
recognizer P systems 119
recurrence 32
regulatory 194
relevant membrane 129
resources 193
rewriting systems 151
RUBAM 205
rule base 208
rule-based 192
rule-based programming 182
rules 193

S

signalling resource 212
situated automaton 247
size 89
skin membrane 6
space-based 217
state 32

state transition 32
statecharts 247
statistical correlations 224
string transition 36
structure-free DNA word sets 65
subset sum problem 115
systems biology 225

T

T cell activation 224
T cell responses 224
T cell signalling network 224
TCR signalling complexes 238
TCR/CD28 240
three-colorability problem 100
time complexity 97
tissue cells 244
topological collection 152
tuple space 217
Turing machine(s) 1, 96, 248

U

usable multiset 136
used multiset 137

V

value function 124
VDM 247
virtual bacterium 207

W

weight function 123

X

X-machines 245

Z

Z 247

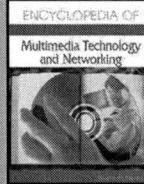